Intellectual Property for Managers and Investors

Technologists have the ideas. Lawyers know the rules. But for business managers and investors, rules and ideas do not readily combine into a strategic vision. No longer is intellectual property (IP) just a necessary expense for large technology companies. Competing and succeding in today's marketplace requires an in-depth understanding of IP – its use as a weapon, as a shield, and as a monetizable asset.

Yet, in a world where fortunes can rise or founder on the strength of an IP portfolio, hesitation to enter this arcane, unfamiliar world still abounds. This book equips the business manager with a working, practical knowledge essential to creating and exploiting IP wealth. It shows investors how to evaluate IP strength and competitive value. With its results-oriented perspective and international focus, *Intellectual Property for Managers and Investors* is essential for those with decision-making responsibility at the interface where business and innovation meet.

Steven J. Frank is a partner in the Boston law firm Goodwin Procter LLP and specializes in intellectual property and business transactions that involve IP. He graduated from Brown University (Sc.B., Chemistry, magna cum laude) and Harvard Law School (J. D., cum laude), and has authored numerous publications involving legal aspects of technology, as well as the legal primer *Learning the Law.* He regularly speaks at conferences relating to law and technology, and is a member of the Sigma Xi Society, and IEEE.

Intellectual Property for Managers and Investors

A Guide to Evaluating, Protecting and Exploiting IP

Steven J. Frank

CAMBRIDGE UNIVERSITY PRESS
Cambridge, New York, Melbourne, Madrid, Cape Town,
Singapore, São Paulo, Delhi, Mexico City

Cambridge University Press
The Edinburgh Building, Cambridge CB2 8RU, UK

Published in the United States of America by
Cambridge University Press, New York

www.cambridge.org
Information on this title: www.cambridge.org/9781107407466

First published 2006
First paperback edition 2012

A catalogue record for this publication is available from the British Library

ISBN 978-0-521-85106-0 Hardback
ISBN 978-1-107-40746-6 Paperback

To Andrea, again and always

Contents

Preface

It wasn't so long ago that intellectual property (IP) didn't much matter to company managers and investors. Copyright was for writers and artists, trade secrets were difficult to protect, and, as for patents, they seemed at best a necessary expense for large technology companies. Not that IP was exactly ignored; epic patent battles followed the introduction of photography, the telegraph, and television, to name just a few disruptive technologies, while the music and broadcast industries struggled over copyright royalties for most of the twentieth century. But few companies viewed IP as a strategic asset. Particularly in the United States, courts historically detected the foul odor of monopoly when asked to enforce IP rights, and the universe of protectible subject matter was limited – software, for example, lay firmly outside the patent system, while the prospects for biotechnology remained uncertain. Lawsuits involving patents were (and still are) enormously expensive, yet few companies licensed their IP routinely and voluntarily. These circumstances left IP uncertain in scope, difficult to enforce, and unlikely to yield an economic return, absent exhausting and potentially ruinous litigation.

As a result, managers spent little time fretting over IP. Often disinclined to dive too deeply into the details of their companies' technology in any case, many simply delegated, trusting that responsibility would reach the sober hands of engineers, scientists, and lawyers padding about their offices in mismatched socks. And why not? Business success seemed to depend no more on IP than the locks on the doors. What more could be done than defining a reasonably secure perimeter around the company's innovation, keeping competitors a step behind if not at bay?

By the 1980s, forces that would dramatically change this static picture began to converge. Commercial software was starting to infiltrate the mass market. Unlike most goods, software is licensed rather than sold so that the originator can retain the underlying IP rights (in particular copyright, which had recently been extended to cover software). Suddenly licensing became a familiar mechanism for disseminating technology, and the notion of sharing innovation with total strangers didn't seem so fraught. Certainly it seemed like a good idea to universities, which, thanks to the Bayh-Dole Act of 1980, found themselves able to keep the rights to inventions made with government funding. In order to bring these inventions into the marketplace – an explicit goal of Bayh-Dole – universities began accumulating patents and licensing them to industry.

It also seemed like a good idea to IBM. For decades one of the world's leading patent recipients, IBM began to license its IP to all comers as a way of generating revenue. To those inured to the notion of patents as a defensive wall – even in the computer industry, these were the days of proprietary architectures – IBM's decision to install a tollbooth seemed heretical. By welcoming smaller, nimbler competitors, wouldn't IBM ensure its own eventual irrelevancy?

Years passed and IBM prospered, its licensing revenues growing impressively. Keen observers warmed to the notion of IP as a monetizable commodity. All that remained was for licensing practices to become institutionalized. Here the telecommunications industry led the way. Think of the telephone network: its value to any particular user derives from the number of other people who also use it. Expansion of a network benefits both consumers and the industry players that create its infrastructure, providing a strong motivation toward standardization – the more the network is based on common designs and technical specifications, the more easily new products can be brought into the network. In response to the rapid growth of telecommunications and the increasing need for compatibility across continents (not to mention the imperative of containing the system's increasing complexity), standards-setting organizations began to proliferate. Soon the benefits of standardization became evident to non-network markets such as the computer industry, the semiconductor industry, and even the highly segmented software industry, as familiarity with common features lured customers and encouraged the development of complementary products. (The auto industry, of course, had learned this lesson generations earlier.)

Technical specifications don't grow on trees, however. While some technology developers may be willing to donate their efforts to a standard in order to fuel marketplace acceptance of their products, others seek to profit directly from their labors. Standards-setting organizations, always seeking the very best technologies, have largely accommodated them. Today most such organizations permit for-profit licensing by their contributors.

At the same time IP licensing was proliferating, its value and scope were increasing. In 1980 the US Supreme Court issued its landmark *Chakrabarty* ruling, which announced a policy of interpreting the patent laws broadly. Sanctioning patent coverage for engineered micro-organisms, the court held that patents could cover "anything under the sun that is made by man." The introduction two years later of a specialized appeals court for patent cases in the United States further signaled a historic reversal of judicial sentiment against patents, and brought much-needed nationwide uniformity to legal standards. The United States accorded copyright protection to software in 1978, Japan did so in 1985, and a European initiative was approved in 1991 (although some member states had already enacted the necessary legislation). With the emergence of clear protection for biotechnology and software, IP law was keeping pace with the fastest-growing areas of industrial innovation, which, in turn, grew ever more dependent on – and creative with – the vehicles IP law provided. Today few businesses (and, as a consequence, few merger partners, acquirers, and equity investors) can afford to ignore them.

Still, IP often strikes fear into the hearts of those it touches due to its sometimes arcane laws and the technical nature of its subject matter. Rather than fear, they should thrill at the sheer range of options and potential strategies now available. IP can be valued, exploited, and traded – even securitized – outside the context of litigation. No longer is there much question about what can be protected. Consider the number of patents issued annually in the United States, which increased from 66,000 in 1981 to 166,000 in 2001. Such growth reflects not only the surging importance of IP, but also the ever-broadening range of enterprises that rely on it.

The aim of this book is to acquaint managers with basic IP concepts, current strategies for its acquisition and exploitation, and how IP strength can be evaluated meaningfully. The term "managers" is intended in the broadest possible sense – research-group leaders, company founders, investors in technology enterprises, corporate sachems and industry mandarins, all-knowing visionaries of every stripe . . . anyone with decision-making responsibility at the interface where business and innovation meet. Corporate and university research managers need cost-effective programs for developing IP that adhere to a sensible budget, while cultivating enthusiasm and cooperation on the part of innovators. A prospective CEO must be able to distinguish a disastrous IP picture from a promising one, and, if he or she decides to join the organization, to define and pursue a realistic strategy. Before investing in a new portfolio company, a venture investor needs an understanding of management's approach to IP and how their efforts have, or will, support business objectives.

IP, in short, forms the bones of this book, but its flesh is all business. In a well-run company, as, it is hoped, in this introduction to the subject, the two are inseparable.

Acknowledgments

For their helpful suggestions and astute observations, I offer special thanks to my colleagues and friends Tom Turano (whose efforts were especially heroic), Dave Byer, Doug Kline, Will Elias, Dave Ting, Karen Copenhaver, Mike Brodowski, Wayne Slater, and Stewart McCuaig. I also thank Mark Beloborodov and Natasha Us for their contributions.

1 Sketching the IP landscape

IP? What's IP?

What's in your head is intellect – the mind, emotions, imagination and creativity, problem-solving ability. "Intellectual property" (IP) establishes rights in intellect made real – writing, art, music, invention. IP rights are bastions of ownership created by law and granted automatically or by government agency or decree. Though intangible, IP is no less real than a bank account or citizenship. But IP rights are creatures of the laws that underpin them. Only those expressions of intellect falling within a favored category receive protection, allowing the owner to prevent unauthorized use. Everything else is unprotectible; it belongs to the public.

Most creative efforts will be eligible for protection within some IP regime. The traditional categories, and the ones with which we will be most concerned, are:

- patents: protect most technologies – useful articles and machines, processes, and compositions of matter, as well as ornamental designs and plants;
- trademarks: protect words, names, symbols, sounds, or colors that distinguish goods and services;
- copyrights: protect works of authorship, such as writings, music, and artworks that have been tangibly expressed, as well as computer software;
- trade secrets: information, such as formulas and manufacturing techniques, that companies keep secret to give them an advantage over their competitors.

It is important to distinguish IP from obligations created by contract, custom, or other law; for example, anything an employee creates in the course of his work may belong to his employer. Some of what he creates may take the form of IP or may be protectible by IP, but the obligation itself stems from the employment relationship and stands outside IP law. It is also important to distinguish the IP right from the fruits of its exploitation. Suppose a battery manufacturer obtains a patent on a long-life battery. A customer buys one.

Do the manufacturer's IP rights end with the sale? Or can it charge the customer additional fees depending on how she uses the battery? The answer, in general, is that IP rights end – are "exhausted" – after the first sale. They do not remain attached to an item, through its life and travels, like an ankle-biting terrier to a pantleg.

Let us consider patents, copyright, and trade secrets in greater detail. These are the IP systems of most immediate interest to technology companies. Because the laws in industrialized countries tend toward similarity more than difference, this book will attempt to avoid excessive focus on the laws of any one place, and instead stick to general principles. Where distinctions matter or if specifics are called for, we will consider the laws of the United States, Europe, and Japan.

Patents

A patent is a government-granted right to stop others from doing certain things specified in the patent document. As such, it is not a monopoly, because it confers no right to actually do anything – just to stop others. Suppose, for example, that you live in an earlier day and your father is the first to invent the automobile. He patents, among other things, the transmission. Grateful for the new mode of transportation, but growing weary of operating the clutch and stickshift as you cruise the boulevards for admiring glances, you invent and patent the *automatic* transmission. Yes, you can stop the old man from making automatic transmissions, you disloyal rascal. But your patent does not give *you* the right to make them, either: if your father's basic transmission patent is broad enough to cover automatic transmissions – that's right, a patent can cover later-developed technology unknown at the time it's filed – his patent "dominates" yours. Neither of you can make automatic transmissions without the other's permission.

Note that the activity, rather than the patents, is the problem here. Patents do not infringe other patents. If no one manufactures automatic transmissions, the patents happily coexist. Only some sort of activity, that is, making, using, or selling something, constitutes infringement.[1] Nor does

[1] Of course, it is possible for the patent office to make a mistake and issue two patents that overlap. In that case coexistence is impossible and the matter must be straightened out. In most countries it is easy – the first to have filed wins as to all subject matter within the scope of his or her patent. In the United States, however, where priority goes to the first to invent rather than the first to file, proceedings in court or before the patent office must be instituted to investigate who invented what first and assign ownership accordingly.

it matter whether that activity was undertaken in ignorance of someone else's patent rights. Independent development is no defense to patent infringement. This is an important distinction between patents and other forms of IP.

Enforceable rights do not arise until the patent "issues" as a formal document. That typically occurs one and a half to three years after the application is filed, depending on the country and the area of technology. In the interim, competitors may freely use the technology, since no IP rights exist yet; the best the applicant can do is declare "patent pending" in the hope of intimidating potential copyists into stealing from someone else. Meanwhile, the patent office examines the application and corresponds with the applicant. Despite this unavoidable period of delay, which can vary widely, the term of a patent is measured from the date the application is first filed. The usual term is 20 years from this date so that, on average, patents offer 16 to 18 years of exclusivity.

Eligibility

A patent applicant must satisfy certain eligibility criteria before his application will even be considered:

- the invention must fall within a statutory category of eligible subject matter;
- pre-filing activity must not have created a legal barrier to patenting.

Although patent laws tend to enumerate specific categories of protectible subject matter, they embrace virtually all technologies.[2] Sometimes, however, specific islands within this sea of coverage remain off-limits. Europe, for example, tends to take a more limited view than the United States when it comes to patenting computer programs (although the law there continues to develop), animal and plant life forms and their methods of production, and methods of treatment of humans or animals for surgery, therapy, or diagnosis. Medical subject matter can be patented in the United States, but patents cannot be enforced against "medical practitioners" performing "medical activities." In other words, while pharmaceutical products and medical devices are fair game, their use cannot be restricted.

In addition, the sea of patentable subject matter is not without shorelines. In most countries, inventions must have a technical character in order to qualify for patent coverage. That is less true in the United States, however,

[2] In the United States, "anything under the sun that is made by man" can be patented. Article 2(1) of the Japanese law defines patentable subject matter as "the highly advanced creation of technical ideas utilizing natural laws."

which has recently become far more permissive toward so-called "business methods." Beginning in 1998, US courts opened the door to patenting of business methods without really defining what they are. The key court decision[3] involved a data-processing system for managing financial services, but, because judges set no limits, recent practice has extended patentability well beyond computer-implemented inventions and even beyond any reasonable notion of "business." Some well-publicized embarrassments – including an unfortunate patent issued for "a method of swinging on a swing"[4] – have prompted greater discretion on the part of the US Patent and Trademark Office (PTO), but the fact remains that, in the United States, little other than laws of nature and perpetual-motion machines falls outside the reach of the patent system.

Patent laws also limit eligibility to subject matter that is genuinely "new." While that may seem painfully obvious, the law considers newness from the viewpoint of the public rather than the inventor. Consequently, an inventor's own activities in selling or calling attention to an invention can preclude her ability to obtain patent protection. The patent system, in other words, is designed as much to safeguard the public as to protect innovation. If the public knows about your invention and you have not sought patent protection, people are entitled to assume you *are not going to* seek protection. Different countries tolerate different amounts of delay. Most, in fact, tolerate essentially none whatsoever. Outside the United States, any sale or public use or disclosure prior to filing a patent application is typically fatal – if you did not file your application before you first sold or publicly divulged your invention, it is already too late. Your application will be denied, or the resulting patent can be overturned if challenged. This rigid rule, obviously, represents a dangerous trap for the unwary.

The United States is more forgiving, allowing applicants a full year to file a patent application following the first public disclosure or *offer* to sell an invention. But note that key difference: the one-year clock starts ticking the moment an invention is held out for sale, so long as it is "ready for patenting" at the time. That does not include licensing, however. An inventor may offer to license an invention without loss of patent rights; but, if the inventor (or his licensee) publicly discloses or commercially exploits the invention itself, he must file within one year.

[3] *The State Street Bank and Trust Company* v. *Signature Financial Group, Inc.*, 149 F.3d 1368, 47 USPQ 2d 1596 (Fed. Cir. 1998).

[4] Think I'm kidding? See US Patent No. 6,368,227, available at www.uspto.gov. You might also have a look at US Patent No. 5,443,036, which covers a "method of exercising a cat" by wiggling the beam of a laser pointer along the floor so kitty gives chase.

This basic tripwire – one year in the United States, no time at all in "strict novelty" countries – is very easily snagged. "Public" disclosures, for example, need not involve the public at all. A nonconfidential discussion with even a single individual, who both understands the technology and is in a position to disseminate the knowledge, can qualify as a public disclosure or use. The trap can be dodged by avoiding pre-filing sales activities and/or entering into suitable confidentiality agreements. Let us consider some typical circumstances that may give rise to a patent-defeating disclosure or sale:

- *Beta agreements.* Companies generally assume that allowing trial ("beta") use of their technology prior to commercial sales falls outside the patent laws. In fact, beta arrangements may well catch the disclosure/sale tripwire if: (i) the originator receives compensation, (ii) the beta arrangement too easily leads to a subsequent sale, and/or (iii) it fails to require confidentiality. In the United States, a limited "experimental use" exception can override the presumption of public use even in the absence of explicit confidentiality requirements; experimentation may be inferred if, for example, the beta site furnishes test results to the originator and returns all materials following the evaluation period. But, while experimental use can trump public use, it *will not* avoid the bar stemming from untimely sales.
- *"Black-box" uses.* What if a company exhibits its next-generation product, still under development, at a trade show? Have patent rights been compromised even if nothing was offered for sale? Perhaps not, depending on what was shown. If there has been no "divulgation" of the invention's operation – for example, viewers merely observe the product's capabilities rather than how it achieves them – then the way the invention works may still be protectible. The damage to future patent rights is limited to what is actually displayed. But, if the mechanism of operation can be inferred from the results, even a "black-box" demonstration can destroy patent rights. Moreover, sometimes recognition of a problem can itself constitute a patentable invention. At the very least, an inventor who publicly demonstrates a solution will be barred from patenting the *concept* of solving the problem, although she may be able to patent the details of her solution.
- *Presentations to prospective investors.* Few professional investors will enter into a confidentiality agreement (at least prior to offering a term sheet). Is a "pitch" meeting with venture capitalists a public disclosure? Maybe or maybe not, depending on the circumstances, but the risk is very much with the pitcher. *Get on file first!*

In addition to qualifying as patentable subject matter and as new in the disclosure/sale sense, an invention must, of course, be new in the technical sense. But it must also be *inventive* to merit a patent. When judged against prior efforts, an invention has to be different in a way that makes a difference – reflecting more than, say, a pedestrian design choice (a rivet rather than a screw) or a trifling variation (a pH of 7.1 instead of 7.2). A patentable innovation, in other words, is a meaningful one. Not necessarily profound – just more than a routine variation or alternative.

Rights protected

A patent owner possesses the right to prevent others from making, using, selling, offering for sale, and importing subject matter that infringes the patent. This raises three questions:

(i) What's infringement? The terms of a patent are highly specific so that the public can know exactly what does, and does not, come within its ambit. Anything that does, infringes. Anything that does not . . . well, it probably does not infringe, but might do. In many countries, including the United States, Japan, and in Europe, the "doctrine of equivalents" extends patents beyond their literal terms to cover subject matter they do not expressly mention. The doctrine is applied sparingly, lest patent claims lose all meaning; courts tend to resort to it when someone clearly obtains the benefits of an invention by departing only slightly from the terms of the patent. Often a whiff of unfair play (for example, contrived readings of ambiguous patent language or a deviation that seems almost cynical in its triviality) is necessary to stir the doctrine into action.

Infringement can be direct – doing what the patent claim says – or indirect. An indirect infringer either "induces" someone else to infringe (for example, by providing how-to instructions and encouraging the infringing activity) or, as a "contributory" infringer, facilitates the violation by providing some enabling component. But that component – a machine that carries out a patented process, for example, or a critical element of a patented device – somehow must be specialized to what is patented and not a "staple" commodity having non-infringing uses. Indirect infringers face the same legal sanction as direct infringers, but only if there is, in fact, a direct infringer out there – if, in other words, the inducement or contributory efforts succeed. The attempt itself is not enough to trigger liability.

(ii) What's the remedy? In general, a patent owner can obtain money damages for past infringement and an injunction – a court order to stop – to prevent

future infringement. Money damages may be based on the patent owner's lost profits or a court's estimation of a reasonable royalty.[5]

Although a patent owner will often consent to continued infringement for the right price, it need not do so. Daft as the decision may be from an economic standpoint, the patent owner usually has the prerogative to preclude anyone or everyone from making his invention. Only in certain cases will he be required to tolerate unwanted use of his IP rights, and even then only for reasonable compensation. In the United States, government funding of the invention's development or an abstract notion of the "public interest" may result in compulsory licensing. Some countries, such as Japan and China, impose a "working" requirement, meaning that, if the patent owner has not commercially exploited the invention within a certain period (typically three years) after grant, others may apply to the government for a license to do so.

(iii) Where is the patent effective? Only within the borders of the country that issued the patent. Suppose you have a United States patent covering a revolutionary toilet valve. That means you can stop any infringing activity having a direct nexus to the United States. US manufacture is covered, even if the valves are intended to be sold abroad. Likewise, importation of foreign-made valves as well as their use in the United States are covered. (Remember, though, that even though unauthorized importation and use are separate offenses, the "first sale" doctrine prevents a patent owner from extracting a royalty from the valve's importer and then from the user.)

United States law takes matters a step further when it comes to foreign activity. Suppose someone ships the toilet valve's individual *components* to Canada and has them assembled there for sale. Since the finished valve is never made or sold in the United States, US patent law would seem to have been outwitted. But no. The law expressly covers such efforts at circumvention, deeming them an infringement as if the assembly had occurred in the United States. Similarly, consider a US patent on a process for making cheese. If someone makes the cheese in Canada but sells it in the United States, that is also an infringement, even though the patent only covers the production process and not the cheese itself; once again, the law applies as if the process

[5] A reasonable royalty is the minimum. Patent owners usually seek lost profits, which can exceed reasonable royalties by a considerable margin. But lost profits are only awarded for profits actually lost, not profits the patentee might have hoped for. To obtain lost profits, the patent owner must prove that there was a demand for the patented product during the period of infringing sales, that there were no acceptable non-infringing substitutes on the market, and that the patent owner had the ability to meet the demand for the products covered by the patent. The patent owner must also provide a detailed computation of the amount of profits it would have obtained had it made the infringer's sales.

had been carried out in the US. More on these exceptions later. For now, think of patents basically as creatures of their home countries. As a result . . .

International rights

. . . applicants seeking protection abroad must apply for patents on a country-by-country basis. The Paris Convention helps make this bearable. A multilateral treaty that has been adopted by virtually all industrialized nations, the Convention assists international applicants by obligating every member country to respect for one year the filing date of a patent application in another member country. Let us say you file a patent application in the United States on January 2, 2006. So long as you file counterpart applications in other countries within one year, they will be treated as if filed on January 2, 2006. This means that disclosures or sales following the United States priority filing are fine; they will not undermine non-US rights so long as foreign applications (or a PCT application[6]) are ultimately filed within a year of the priority date (see figure 1.1).

Do not confuse this one-year priority-hold period with the one-year disclosure/sale grace period accorded in the United States. Most every country will respect a priority date for one year; only the United States allows you to *delay* securing a priority date for up to one year after a disclosure or offer for sale. So, if you only care about rights in the United States, go ahead and disclose or sell your invention to your heart's content; just file within a year of when you start. If you want to preserve rights elsewhere, you must file *somewhere* before any public disclosures or sales; then you get a year to file foreign counterparts.

Since 1995, United States patent applicants have had another option. A "provisional" patent application is a foot in the door. It need not contain claims or have any particular organization or content. Within a year of the provisional filing date, however, a more complete, garden-variety "non-provisional" patent application must be filed. The one-year priority-hold period for foreign applications also begins at the filing date of the *provisional* application. Accordingly, in addition to filing the US non-provisional, the applicant must also file any non-US counterpart applications by the first anniversary of the provisional filing. Other countries, such as the United Kingdom, also permit filing

[6] More on Patent Cooperation Treaty (PCT) applications below. For now, think of them as placeholders that allow you to defer filing of foreign counterparts – which can be expensive – for up to an additional 18 or 19 months while preserving the original priority date.

of provisional applications (although they may be called something different), with identical timing requirements for domestic and foreign follow-up.

For priority purposes, a provisional application is no different from a non-provisional. If filed within a year of a public disclosure or offer for sale, the provisional application theoretically preserves United States (but not foreign!) patent rights, according the applicant an additional year before the final US application must be filed. The provisional also triggers the one-year Paris Convention priority hold for foreign applications, so filing before sales or disclosures theoretically preserves non-US rights as well. But the qualifier in both cases is "theoretically." The reason is that the provisional is only as good as what it describes. Make a later patent claim that is not supported by adequate teaching in the provisional, and you can forget about the provisional's priority date – both in the United States and abroad.

This is why patent lawyers hate provisionals. Clients often assume they can make do with an inexpensive, stripped-down provisional, and lawyers who advise the fully loaded non-provisional model are just playing salesmen. But, too often provisionals offer a false sense of security. To develop confidence in the sufficiency of any patent application, a patent attorney must learn about the invention, consider possible workarounds, and satisfy herself that the application teaches how to make and use everything the inventor wants to cover. Anything less and the application fails. So, while the patent laws allow you to slap a cover sheet on a Ph.D. thesis or on the PowerPoint presentation you prepared for venture capitalists and call the result a provisional application, it is impossible to know, until the real work of a patent application is done, whether it will stand up.

Still awake? Then prepare yourself for the final international complication: foreign-filing licenses. Some countries, including the United States, China, the Russian Federation, and various European states,[7] require applicants to obtain a license from the patent office before applying for a patent in any foreign country; this gives the government the chance to consider the national-security implications of the application and, if necessary, issue a secrecy order that may suppress the application indefinitely.[8] Other countries, including Japan, Canada, and various (different) European states, have no such restrictions. In the United States the requirement depends on the place where invention

[7] France, Italy, Poland, and the United Kingdom impose restrictions to varying degrees.

[8] Few patents are the subject of secrecy orders, but the standards by which the occasional order is issued and its longevity can vary widely. A 1958 patent application filed by three US Army chemists for a method of synthesizing the deadly nerve agent VX was understandably suppressed by the US PTO, but less understandably declassified in 1975. It is now publicly available.

occurs rather than the nationality of the applicant; anyone who makes an invention in the US must obtain a foreign-filing license before filing abroad – even in his home country.

The penalties for filing abroad without a license vary. In the United States, any resulting United States patent is invalid unless the PTO, in its discretion, issues a license retroactively. In France the inventor can land in jail.

The need for a foreign-filing license can raise blood pressures as the one-year anniversary of the original priority application approaches. Suppose that a US applicant has a provisional application on file, but waits nearly the full year before filing the non-provisional. It may be months before the PTO gets around to issuing the foreign-filing license for the nonprovisional, and, in the meantime, filing abroad is impermissible. Even if the original provisional application received a suitable license, it may not cover the nonprovisional. Let's say the deadline arrives before the license – what now? The applicant must either forgo foreign filing altogether or file a PCT application with the PTO (thereby giving the US government the opportunity to consider a secrecy order) to preserve foreign rights.

The mechanics

Fear not, we won't dig very deeply into the eye-glazing arcana of patent procurement – just enough to provide a sense of requirements, timing, and cost. From the perspective of the inventor, the process of obtaining a patent goes something like this: you work with your patent lawyer or agent to prepare an application that describes your invention in exhaustive (and expensive) detail; the finest years of your life slip away as the application languishes, awaiting examination; at last the patent office awakens, only to curtly reject all of your claims, seeming almost astonished at your nerve; your lawyer says not to panic and, in most cases, eventually persuades the patent examiner to allow at least some claims; and, finally, the stiff-covered patent award issues forth.

Let us look a little more closely at the patent application and how it is prepared. The grand bargain underlying the patent system is this: *sow knowledge and ye shall reap the rewards of exclusivity*. In exchange for educating the public about your invention – in fact, teaching them in sufficient detail that anyone with a reasonable amount of skill can, after reading your patent, make and use the invention – you receive a limited period of exclusive rights. Yet, despite the restrictions that the owner of any one patent may impose, the system of patents is designed to enhance, not limit, innovation. Patents are stepping stones. Free to absorb the ideas they teach, the public is not merely allowed but *expected* to innovate beyond those ideas; avoiding patents by improving upon them is bad news for the patent owner, but a great boon to progress.

All of this means that, for the system to function properly, the disclosure in a patent must be comprehensive. We will consider the specific requirements later. For now, suffice it to say that the patent application is an extensive teaching document and, as such, can be pricey. The average cost ranges from $5,000 to $20,000, depending on complexity, the number of discrete inventions involved, the preparer's familiarity with the subject matter, and the willingness of the inventor to shoulder part of the preparation burden. The bulk of the application is devoted to a description of the invention, and, at the end, the invention is painstakingly defined – often in excruciating and obscurely phrased detail – in a series of patent claims. Think of the patent claims as a property deed. They must, in words, somehow capture the essentials of an invention just as a deed uses words to map territorial boundaries. Patent claims make difficult reading because, first, language can be a clumsy and imperfect tool for expressing abstract concepts;[9] it is a lot easier to point to the beginning and end of a lot than it is to describe property partitions and their relationships on paper. And, second, as if the limits of language were not already stretched to capture the abstract, the law also requires great precision. Patent claims, again like a deed, must warn the public where it may not tread. The contradictory demands of abstraction and precision lead to bizarre verbal locutions that torment the normal mind.

Patent claims are arranged hierarchically, with successive claims adding extra elements. These define strategic retreat positions. Each claim stands on its own, and each must define inventive subject matter. If the broader claims are successfully attacked in litigation, perhaps the narrower ones will survive – and the patent owner will be entitled to the same remedies whether one claim or a hundred are infringed.

Given the quid pro quo of teaching for protection, one might expect that a patent application would remain secret until the day the patent issues, affording its owner the opportunity to decide whether the coverage ultimately offered merits the disclosure required. One would be wrong, however. Almost without exception, patent offices publish applications 18 months after filing. At that

[9] Poppycock, says the skeptical reader: patent claims involve cams and chemicals and circuits, all entirely concrete items. Where is the abstraction? The answer is that an invention is not a disembodied aggregation of parts, but typically an objective – liftoff, a message transmitted, a disease cured – that is achieved in a particular way. Perhaps some realizations of the invention rely on certain parts to make that objective happen. But no inventor wants patent protection to be limited to a single implementation. Framing a description of the objective and how it is achieved without being limited to a specific set of parts is the essence, and the challenge, of a good patent claim.

point, the invention can no longer be maintained as a trade secret. What's worse, publication usually takes place before a patent examiner has even got round to reviewing the application. Unless a patent ultimately issues with claims mirroring what's taught, the world will have a free how-to manual. Wasted time and money, in other words, may be the smallest penalties inflicted by a doomed patent application.

The only limited exception to this mandatory soul-baring occurs in the United States. Although automatic publication was adopted in 2000, as a nod to historical practice, US patent law still permits the applicant to prevent publication – so long as she states an intention not to file abroad (where publication would be automatic). Seldom does anyone avail herself of this option. For one thing, the foreign-filing decision is usually deferred until the end of the one-year priority-hold period, whereas the request not to publish must be made upon filing of the US non-provisional application. More importantly, sacrificing foreign rights prematurely can prove costly in the long run.

Then again, actually filing in foreign countries is guaranteed to be costly in the short run. Even hardbitten managers may curl into a fetal position when confronted with the fees and expenses. Some countries treat their patent offices as profit centers, charging applicants accordingly, and it can seem that every incremental step in the process – filing, requesting examination (oh, you wanted us to *review* that application you just paid a fortune to file?), merely keeping the application in force year to year – involves payment of a fee. The need to translate the application into foreign languages further augments the high up-front filing price tag. Still, the deadline arrives a mere 12 months after the priority date. How often is an undeveloped product's potential known at such an early stage, much less known well enough to gauge the strength of foreign markets against the cost of protection there?

This is where the Patent Cooperation Treaty (PCT) comes in. The PCT, in essence, affords the opportunity for further delay. An "international" application filed under the PCT is a placeholder, or, perhaps more accurately, a call option. For $2000–5000, depending on various factors, a PCT application allows the actual filing of foreign applications to be deferred for an extra year and a half (for a total of 30 months – see figure 1.2) without loss of the original priority date. Moreover, 18 months after the priority filing, the application is published with an "international search report" performed by a competent patent authority (often the European or US patent office). The search report provides the applicant with an initial sense of how his claims stack up against the prior literature. Thus, when the 30-month deadline actually rolls around, he can make an informed foreign-filing decision based on a forecast of the patent protection likely to ensue as well as, hopefully, a growing understanding of the product's global potential.

The price of all this delay, of course, extends well beyond the modest PCT filing fee. The application does not even get in line for examination until the foreign applications are filed, and meanwhile the 20-year patent term, measured from the date of the PCT filing, ebbs ever so tragically away.

Examination, when it eventually occurs, ushers in a period of haggling. Rare is the application that is allowed immediately. (Indeed, should it happen, the patent lawyer may feel a queasy sense of not having asked for enough going in.) The process of haggling, called prosecution, can take a long time: months, usually, but sometimes years. And, if ultimately unsuccessful, the patent applicant may choose to appeal the examiner's rejections – a process that will itself take years. All of this as the 20-year term clock ticks on. It is for this reason that few applications are appealed. Patent lawyers fervently attempt to reach some accommodation with the examiner during the prosecution stage.

Before or after a patent is granted, depending on the country, third parties may have the opportunity to oppose it. Such proceedings, once again, add time and cost to the process, since nothing limits the number of potential challengers.

These examination procedures do not vary much across countries except in terms of timing and formalities. India, for example, has a notorious backlog of patent applications that can delay grant for more than a decade (as the patent term, once again, bleeds away). Some countries, such as Mexico and Brazil, essentially defer to the outcome of proceedings in the United States or Europe. Other countries, such as Canada, do not expressly defer, but take an inordinate interest in such proceedings. Still other countries – various Middle Eastern states, for example – do not examine patent applications for anything beyond formal compliance; these "registration" countries simply publish patents as filed, and allow the courts to determine proper coverage should the owner sue anyone.

The most important procedural departures occur in countries organized into regions, notably Europe, where a single proceeding governs patent protection in a group of countries. The successful patent applicant in Europe obtains a patent grant from the European Patent Office in Munich, Germany. But this patent is not, in itself, enforceable anywhere. Instead, it must be "validated" in each European state where it is to be enforced. Significantly, each such country is free to interpret and enforce the patent in accordance with its own laws. There is, therefore, no unified law in Europe relating to enforcement and infringement of patents.

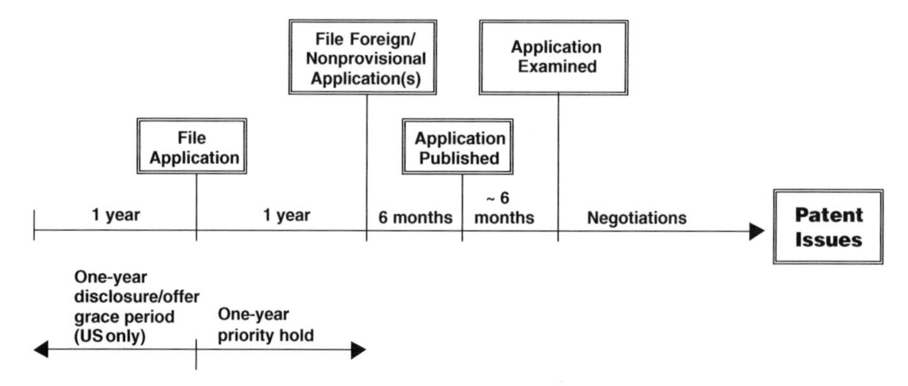

Figure 1.1 U.S. patent prosecution timeline.

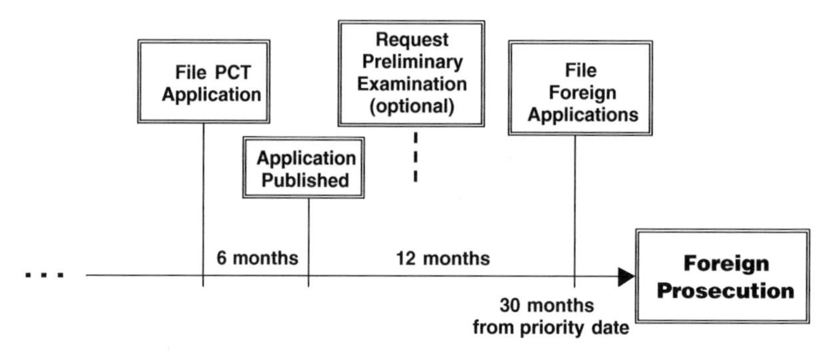

Figure 1.2 International patent application timeline.

In summary, the advantages of patents include:

- *coverage is clear and readily enforced,*
- *broadest range of eligible subject matter,*
- *independent development is no defense to infringement,*

while the primary disadvantages involve:

- *high cost,*
- *long elapsed time between application and grant,*
- *likelihood of disclosure prior to effectiveness of patent.*

Copyrights

If that furious slog through the patent heath has left you breathless, welcome to the obliging oasis of copyright. Here life is so much simpler. Difficult procedures? None – coverage is automatic. International protection? Automatic as well. Eligibility restrictions? Minimal. Cost? Essentially zero. Duration? Almost forever.

Sound too good to be true? It depends, as you might expect, on what you are protecting and how much you hope for. Copyright was devised primarily to protect authors. Writers have always been an ornery bunch, and the invention of the printing press opened the possibility not only of mass distribution, but also of mass theft. The British parliament passed the first copyright law, the Statute of Anne, in 1710, in response to the proclivity of "Printers, Booksellers, and other Persons" to reprint written works without their authors' permissions – "to their very great Detriment, and too often to the Ruin of them and their Families." The statute introduced the idea that an author should own the right to reproduce his work, as well as the principle of a fixed term of protection (a mere 21 years). As copyright law evolved, the term of protection and the kinds of work within its reach grew substantially. No longer limited to literary works, copyright now covers works of art, musical and dramatic compositions and performances, sound and film recordings, radio and television broadcasts, and, most problematically, computer software.

Despite its broad repertoire, copyright ultimately remains bound to its motivational roots – the ease with which creative effort can be reproduced, and thereby stolen, mechanically (or, today, electronically). The core and simplest offense with which copyright is concerned is slavish, shameless copying. Copyright law does a good job, therefore, banning sales of illicit musical recordings, unauthorized discs or downloads of commercial software, and bootleg reprints of books and magazines. Obvious rip-offs fall squarely within its sights.

But beyond obvious rip-offs . . . well, for now let us not cavil. Instead we shall bask in our oasis of user-friendly procedures and laid-back eligibility requirements.

Eligibility

The subject matter of copyright is, as described above, very broad. If something reflects creative expression and can be copied, there is a good chance it comes within the copyright system. Purely utilitarian works, on the other hand, lie outside its scope; that is what patents are for. Some artifacts possess both functional and creative components – for example, decorative lamp bases. In such cases, copyright protects the creative portions but does not stop the rest of the world from making lamp bases.

Beyond qualifying subject matter, the eligibility criteria for copyright are minimal:

- originality,
- fixation in a tangible medium of expression.

For both of these requirements, the bar is set low. "Originality" means that a work has been created independently and reflects at least a smidgeon of intellectual effort. But it need not be original in the sense of novelty or ingenuity. "Fixation" means that the work has been recorded in some concrete way – on paper, on tape, or in a computer memory, for example.

The originality requirement, while modest, is not non-existent. Information databases, such as telephone directories, that involve no creative choice in arrangement or selection lie outside copyright. That has prompted proprietors of large databases to seek special legislation protecting their labors. Their most significant success has occurred in the European Union, which passed a directive mandating 15-year protection for databases, provided that creation of the database required a "substantial investment."

Rights protected

The owner of a copyright receives a bundle of exclusive rights, not all of which are meaningful in every situation. The bundle includes the rights to reproduce or copy, adapt (that is, prepare "derivative" works such as translations or integrations of one computer program into another), distribute, and publicly perform the work. These rights enjoy great longevity. In the United States and Europe, the copyright on a recent work lasts for the lifetime of the author plus 70 years; in most other countries (including Japan) the term is life plus 50 years. US law prescribes a different term for corporate works: the lesser of 95 years from publication or 120 years from creation. Once copyright expires the work enters the public domain and may be copied freely.

Many countries, especially in Europe, also recognize "moral rights," which protect the creator of a work rather than just the work itself. Traditional moral rights include the right of "paternity" (to be identified as the author of a work), the right of "integrity" (to prevent others from distorting the work), and the right of withdrawal (to retract, modify, or disavow the work). Although these rights clearly reflect concern for artists and authors, and seem awkwardly out of place in the world of technology and software, they may nonetheless apply. Moreover, to save starving artists and desperate writers from themselves, moral rights usually cannot be waived – that is, blanket agreements to limit or transfer these rights typically are unenforceable, and the author's consent must be obtained on a case-by-case basis. Without such consent, a software developer who has assigned copyright ownership to his customer may nonetheless rely on moral rights to prevent the customer from making future changes to the software without the developer's express consent.[10]

Over its history, copyright has readily adapted to a wide range of new technologies and media, consistently outlasting those who have declared it obsolete. The near ubiquity of the Internet has posed the latest challenge. Works represented in digital form have never been more readily available, nor have the opportunities for massive, essentially costless duplication and distribution been so widespread. Publishers fret that a single stray digital copy of a popular book or hit song will destroy the legitimate market in an online minute, while legions of file-swappers slow their campus networks to a crawl in an effort to prove them right. Despite all the sky-is-falling dread, however, the Internet has not changed the basic parameters of copyright – the same kinds of works, activities, and livelihoods remain at stake. Instead, what have changed are the medium of dissemination and the players involved, as well as the measures taken by publishers to thwart thievery.

Enter the US Digital Millennium Coypright Act (DMCA), which has, among other things, clarified the liability of the online service providers who plug their subscribers into the Internet, and made life difficult for those who try to circumvent copy-protection measures used to guard against improper duplication. Europe is heading down a similar path. Essentially, the DMCA protects Internet service providers from copyright infringement liability simply for transmitting information, so long as they remove infringing material from users' web sites upon notification. The DMCA also makes it a crime to hack through (that is, circumvent) anti-piracy devices, although it is fine to do so in order to conduct encryption research, assess product interoperability, and test computer security systems. While the DMCA has attracted a firestorm of criticism, some justified and some not, it has not expanded the notion of piracy or suddenly foreclosed activities that previously had been permitted.

International rights

Copyright protection is not only self-executing, but international in scope. A web of treaties, most notably the Berne Convention and the Universal

[10] Moral rights vary dramatically across countries. In the United Kingdom, moral rights can be waived on a blanket basis, and do not cover software. Canadian moral rights cover software but blanket waivers are permitted. In Australia, moral rights cover software and can only be waived on a case-by-case basis for particular uses.

Copyright Convention, bind together the copyright laws of most of the world's countries. Protection in one country automatically extends to others via these treaties.

The basic principle behind them is "national treatment," meaning that each country extends the same protection to citizens of other countries as it does to its own citizens. If the copyright term is longer in the United States than in Japan, for example, a US citizen will receive the shorter term in Japan, but a Japanese cititzen will benefit from the longer term in the United States. The concept of national treatment extends to the scope and character of protection as well – for example, the degree to which software is protected under copyright. With the exception of moral rights, however, most countries base copyright protection on similar principles, so the subject matter and scope of copyright do not vary significantly.

The mechanics

Mercifully few mechanics complicate life in the copyright oasis. Once a qualifying work has been "fixed" in a concrete medium, the bundle of rights attaches without further ado.

The United States, however, values registration of copyrighted works and the deposit of copies in the Library of Congress. Treaty obligations limit the degree to which registration and deposit can be forced, but US law provides strong incentives to undertake these minimal steps. The right to sue, for example: for works created in the United States, the copyright owner cannot file a lawsuit until the work has been registered. Moreover, registration carries certain procedural benefits. If it occurs before the first infringement, the copyright owner may recover "statutory damages" and attorneys' fees.[11] Statutory damages provide a floor on the amount of money that can be recovered in court for a copyright violation, and, depending on the circumstances, that floor can range from $200 to $150,000. The practical effect of statutory damages and attorneys' fees is to make lawsuits viable in cases where the actual monetary harm is small.[12]

We have all seen copyright notices on everything from advertisements to Adobe Acrobat. Such notices are largely optional, but because different

[11] To complicate things a bit, in the case of works published for the first time, registration may take place up to three months after publication without loss of benefits (that is, regardless of when infringement occurs).

[12] In the United States, the usual measure of damages is based on the harm actually suffered as a result of the infringement, and any profits of the infringer not taken into account in computing actual harm.

countries (and treaties) have different rules – in the United States, for example, the presence of a notice is optional but deprives copiers of the ability to claim "innocent" infringement – it is best to add a copyright legend to published works. In 1511, the German artist Albrecht Dürer used this one:

> Hold! You crafty ones, strangers to work, and pilferers of other men's brains. Think not rashly to lay your thievish hands upon my works. Beware! Know you not that I have a grant from the most glorious Emperor Maximillian, that not one throughout the imperial dominion shall be allowed to print or sell fictitious imitations of these engravings? Listen! And bear in mind that if you do so, through spite or through covetousness, not only will your goods be confiscated, but your bodies also placed in mortal danger.

A less operatic notice will suffice today. One that satisfies the major treaties takes the form:

Copyright © 2005 ABC Corporation
All rights reserved.

Limitations of copyright

As noted previously, the rights bestowed by copyright are ideally suited to preventing unauthorized duplication. The less slavish the copying, however, the less coverage copyright will afford. Copyright protects the expression of an idea rather than the idea itself – the Mona Lisa, not the artist's prerogative to paint sloe-eyed, enigmatic beauties. Borrowing is not the same as stealing, in other words, and the gears of copyright begin to creak when engaged to defend "non-literal" aspects of a work.

Where does copying end and permissible borrowing begin? Obviously, if infringement actions could be won only against those who copy slavishly, works could be appropriated through trivial additions or deletions. A common standard of comparison for infringement, therefore, has been defined as "substantial similarity" rather than complete identity. How much similarity qualifies as "substantial" is an elusive question, though, and there is no easy way to judge how close will turn out to be too close under the law.

Perhaps more importantly, substantial similarity by itself is not enough for copyright infringement. Independent creation is always a defense. Copyright, once again, protects against *copying*. If the competitor of a software publisher can show that it created even a completely identical work on its own, copyright will not help the originating publisher. Now, its critics not withstanding, the law is usually no fool, and courts recognize that few defendants cheerfully admit to deliberate copying. In general, it is enough to show that the maker of a substantially similar work had *access* to the original. That is why software developers often go to great lengths to develop competitive code in a "clean room" environment, in effect documenting the absence of access.

If expression is protected but ideas are fair game for all, what happens if there are very few ways to express an idea? How many ways, for example, can you diagram assembly instructions for a bookcase or express the idea $E = mc^2$? In such cases, idea and expression are said to "merge," and copyright does not apply at all (lest ideas be monopolized). The concept of merger has great importance in the context of computer software for two reasons: first, the issue of how many ways exist to code something is highly technical and frequently open to dispute; and, second, it means that the more valuable protection would be – that is, the more difficult it would be for competitors to avoid, given the dearth of alternatives – the less likely that protection is to be available.

A limitation related to merger occurs when the expression is a "stock" feature or incident (a *scène à faire* in copyright parlance) commonly employed in the treatment of a given idea – for example, scenes of soldiers marching in a war movie. No one can have a monopoly in such tropes. Once again, this concept looms particularly large in the software context. If a particular motif has become common, it is not protectible; a copyright case regarding computer displays, for example, held the use of overlapping windows to be a stock aspect of windowed displays and therefore outside copyright. Moreover, if a form of coding is dictated by external factors such as hardware compatibility requirements and industry practices, it will also fall within the *scènes à faire* doctrine.

As we will see later, the effect of all these cutbacks on copyright protection for software can be significant indeed.

Finally, the concept of "fair use" protects certain kinds activities that would otherwise infringe copyright. The idea behind fair use is to protect copying performed for a limited and "transformative" purpose, such as to comment upon, criticize, or parody a copyrighted work. Such things can usually be done without permission from the copyright owner. Whether a given use will be exonerated as a "fair" one depends on various factors, such as the *character* of the use (for example, commercial as opposed to purely educational or public-interest purposes), the *nature* of the work (the more creative it is, the more protection it receives against copying), the amount of the work actually copied, and the effect on the market for the work (even a commercially motivated parody, for example, will not likely steal sales of the original[13]).

In a technology context, fair use comes into play most often in connection with reverse engineering of computer software. Copyright law encourages exploration of a work's underlying ideas. But what happens when getting at those ideas requires unauthorized copying? The issue has often arisen in the context of videogames, specifically attempts to design game cartridges compatible with a particular manufacturer's console. That typically requires figuring out how the console interacts with a cartridge. If the only way to do that is to copy some code (because, for example, the console manufacturer refuses to make compatibility information available), then such copying will probably qualify as fair use.

So, while the benefits of copyright include:

- *low cost, absence of procedural requirements,*
- *automatic international reach,*
- *low standard to qualify*;

the downsides:

- *focus on actual copying,*
- *numerous exclusions limiting scope of protection,*
- *independent development as a defense to infringement*

can be appreciable.

Trade secrets

If patents are the reluctant inamorata whose favors are expensively beseeched and at last parceled out warily, and if copyright is the alluring vixen who promises the world but finally delivers little, then trade secrets are a stern uncle who admonishes you to stop begging the government for help and start fending for yourself.

The patent and, to a lesser extent, copyright laws encourage disclosure, bestowing protection as a means of ensuring the free exchange of ideas and information. Trade secret law performs quite the opposite function: it assists active efforts to maintain confidentiality. The law of trade secrecy is quite unconcerned with protecting innovation. The goal, instead, is to enforce norms

[13] Although a scathing parody might negatively affect sales of the parodied work, that is not a copyright concern.

of commercial conduct and prevent unfair competition. Parties entering into confidential business relationships must remain faithful to promises of nondisclosure. Outsiders may not purloin the expense and efforts of others to attain competitive advantage.

Courts therefore assist those who help themselves. When harmed by unfair conduct, the trade-secret owner can often obtain an injunction – which prevents the wrongful user from exploiting the secret – as well as monetary damages; but only if adequate safeguards have been put into place. The degree to which the law will aid a trade-secret owner depends on the interplay between the precautions she has taken to prevent theft and the manner in which an intrusion occurs. Insufficient measures to guard against theft forfeit protection, regardless of the thief's knavery. Minor leakage in the face of adequate safeguards, on the other hand, only weaken protection: procurement by wrongful means may still be actionable, although the innocent finder of leaked information will not face punishment. Injunctions, when issued, typically run for at least the length of time that would be required to develop the device or process without access to the trade secret (the reverse-engineering period), and possibly longer if the defendant's conduct seems especially scurrilous.

The law of trade secrecy is highly local in nature, varying from country to country and, in the United States, from state to state. The basic scope of what is protected and the remedies available, however, vary little. A trade secret is information (such as a formula, computer program, manufacturing technique, customer list . . . you get the idea) that has value by virtue of its secrecy – that is, the fact that it is not generally known or easily ascertainable by proper means. Computer software qualifies when sufficient originality exists to produce a competitive advantage.

What protection measures should a trade-secret owner take? That depends, of course, on the nature of her business and the likely leakage pathways. Because a court inevitably assesses in hindsight the adequacy of the measures taken, the business owner must always think ahead. Manufacturers worry about key supervisors, and software developers fear misappropriation by programmers, but trade secrets can also be lost to customers, financial sponsors, joint venturers, and competitors who resort to industrial espionage. Some typical precautions include:

- strict access controls;
- site security;
- employee nondisclosure agreements;
- rigid exit interviews;

- contractual restrictions with vendors and customers, including liberal use of nondisclosure agreements at the earliest possible stage of a relationship;
- clear notice, such as conspicuous legends declaring proprietary materials "confidential."

Ultimately, however, the ability to enforce trade secrets depends less on satisfying eligibility criteria than on the practical demands of litigation – the intrinsic hurdles of evidence and procedure that a litigant must overcome to make his case. While such hurdles confront every courtroom supplicant, they loom particularly large against the trade-secret owner. We will see why in the next chapter.

Summarizing, the advantages of trade secrecy include:

- *broad range of qualifying subject matter,*
- *absence of formal requirements;*

while the disadvantages center around:

- *difficulty in enforcement,*
- *defeat by independent development, reverse engineering,*
- *potentially high cost of maintaining secrecy,*
- *varying laws.*

The miscellany

Design patents

Design patents and registrations occupy an uncomfortable middle ground between ordinary (the proper term is "utility") patents and copyright. Design patents cover the ornamental features of items such as lighting fixtures and toys, which, as "useful articles," fall outside the scope of copyright. Functional attributes, on the other hand, lie outside the scope of a design patent; that is why we have utility patents. In the United States the test of infringement for design patents is similarity in the eyes of an "ordinary observer."

If that were the end of the story, design patents would occupy a forgotten little backwater of IP law and no one would pay much attention. But, in fact, people have been (mis)using design patents for years to cover functional, industrial designs like automobile spare parts and mechanical interfaces. Why? Well, sometimes because the product may not qualify for a utility patent, and sometimes to save money: design patents tend to cost far less than their utility cousins. The extent to which this strategy can succeed is, to say the least, questionable – can you really separate the aesthetic attributes of an exhaust manifold from its functional character? – and the wise industrial IP consumer treats design patents as an unreliable but inexpensive fallback position.

Mask works

The US Semiconductor Chip Protection Act is a part of the copyright statute that specifically protects chip designs – in particular, the pattern "masks" used to fabricate microcircuitry. The effect of the law is to protect, for ten years, the three-dimensional pattern or typography of a semiconductor chip produced by photographic processes. The mask work owner has exclusive control, not only over reproduction of the mask work itself, but also over the import or distribution of semiconductor chips incorporating the work.

The Chip Act does not prohibit reverse engineering, however, and competitors may incorporate whatever they legitimately discover into their own mask works so long as these meet copyright's originality requirement. In other words, anyone can duplicate a chip's functionality so long as he uses a different circuit layout. Consequently, the Chip Act is not terribly important even within the narrow community of semiconductor manufacturers who benefit from its terms. The mere fact that Intel's latest microprocessor designs are almost immediately duplicated, in terms of functionality, by competitors illustrates its limitations. Most semiconductor companies rely on patents to protect their significant innovations.

2 Making the strategic choice

The last chapter outlined the various options for protecting innovation, their advantages and disadvantages. If managers could consult these criteria to identify the best strategic options for a given situation as easily as choosing the right fence to surround a yard, IP planning might be a simple matter of comparing price, availability, and preference. Would that it were so easy. Not only is the exercise of comparison more subtle, but, in fact, we have so far considered only half the story. The criteria discussed in chapter 1 set the basic legal parameters – the boundaries of your property. The ability to enforce IP rights, however, means the power to eject squatters. The fanciest estate quickly becomes uninhabitable if anyone can pull in and stay for free. As Odysseus learned when he returned from Troy to a houseful of gluttonous freeloaders with designs on his wife, having a legal property right isn't the same as vindicating it.

Not that anyone should be too eager to resort to litigation. The best lawyers would rather persuade the squatters to leave than smite them as did Odysseus. But it is essential to assess the practical availability of a legal remedy along with the suitability of the property right, since only rights you can enforce – in terms of money, time, and reliability – are worth procuring.

In this chapter we will consider how to evaluate protection mechanisms in real-world situations, focusing first on the context of enforcement, then examining the options through the lens of a case study.

The context – IP dispute resolution

US lawyers see themselves as zealous pursuers of the truth and their clients' interests, which hopefully coincide. That zeal often strikes others as overbearing, even thuggish, turning litigation into a ferocious Darwinian struggle between desperate opponents. And nowhere are opponents more desperate than in the throes of IP litigation, in which each side risks losing its most dearly held assets – the technological crown jewels, wellspring of its success, and, in a very real sense, the source of its identity. Competing views of litigation may never be reconciled, but it is undeniable that US procedure offers more tools and opportunities to uncover the truth (and, unfortunately, more opportunities for abuse) than most other systems.

One reason for this is the inherently adversarial nature of United States law. Like other "common law" countries, such as the United Kingdom, Ireland, and Australia, the US relies on two self-motivated opponents to clash mightily and make their best cases to a judge or a jury, who will sit back and listen and then render a decision. The court's largely passive role leaves it to the parties to root out the relevant facts through a system called pre-trial discovery. In "civil law" countries, such as other European states and Japan, by contrast, judges tend to take a much more active role in the proceedings, often questioning witnesses themselves.[1] Discovery is limited or nonexistent; indeed, even cross examination by lawyers may not be allowed. Obtaining evidence from an adversary's files or premises can be next to impossible. The absence of pre-trial discovery, while perhaps shielding some wrongdoing from detection, nonetheless can dramatically shorten the course of a lawsuit and lower its cost. (Only in Japan, it seems, are trials seemingly endless despite minimal discovery.)

To get a sense of what is involved, let us consider the progress of a typical patent lawsuit in the United States. The patent owner files a complaint. His adversary, the accused infringer, has 20 days to answer. The court sets a litigation schedule, and, over the ensuing 9 months or so, both sides engage in various forms of discovery in a determined hunt for relevant information. Witnesses are *deposed*, that is, formally interviewed under oath before a court reporter. Documents and items such as product samples are requested (sometimes in response to the answers given in depositions). Each side presents a series of "interrogatories" – written questions demanding written answers from the other side. Expert witnesses disclose their opinions and submit to interviews. If one side refuses to comply or drags its feet excessively, its opponent may ask the court to intervene. Similarly, if lawyers become too aggressive in discovery, making excessive demands or pursuing irrelevant avenues, the court may impose sanctions. Still, even when conducted within the bounds

[1] It is unclear why common-law countries tend to have adversarial trial systems, while the systems in civil-law countries are usually "inquisitorial" by nature. The term "common law" refers to the binding nature of prior judicial decisions and the ability of judges to shape the law. A civil-law system, by contrast, relies primarily on statutes and regulations rather than judges' interpretations of them. While earlier decisions are not ignored, they lack the formal weight they receive in common-law countries.

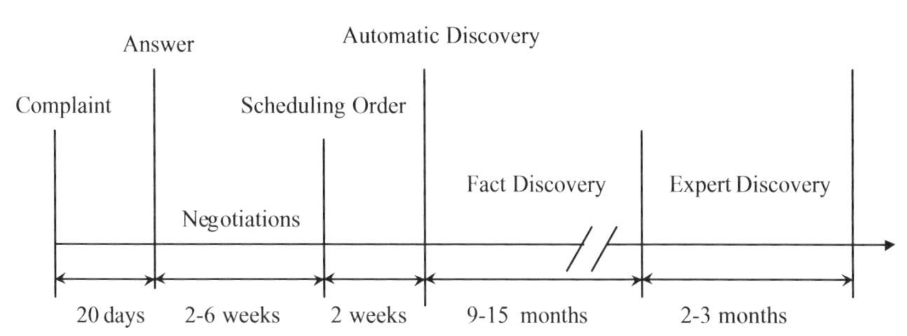

Figure 2.1

of propriety, discovery can be draining and extremely expensive. It may be necessary to locate and turn over thousands of documents, visit witnesses scattered across the country and sometimes beyond, scour electronic records going back years – all while complying with procedures that protect trade secrets, all while busily reviewing the flood of material issuing from the other side (see figure 2.1).

Fact discovery leaves few opportunities for courtroom surprise. At the end of the process, both sides have a much better sense of their cases' strength, often leading to settlement.[2] Lawsuits go to trial when, despite discovery, the two sides cannot agree on the facts (or a settlement). Where the sides do not dispute what happened but disagree on the law (or, more commonly, how it applies), the case may be decided on *summary judgment* – a preliminary short-circuiting procedure that resolves some or all issues short of trial. Should a trial eventually ensue, however, the proceedings can last up to a month, with expensive teams of lawyers and expert witnesses playing to the audience – the United States stands alone in permitting patent cases to be tried before juries – and billing at full bore. Overall, the average duration of a patent lawsuit through trial is two to three years. Add another year for an appeal. Each side's costs can range from from $1 to $2.5 million or more (see figure 2.2).

Is all this heavy legal machinery really necessary to bring out the merits of a case? Not always. In cases of patent infringement involving an easily obtainable and readily understood product, there are no secrets to be unearthed: the patent sets forth the rights involved, and the product either does or does

[2] In European countries, particularly Germany, settlement is relatively rare; infringement proceedings are so expeditious (6 months from start to finish is not uncommon) and the overall cost so small a fraction of that faced by a US litigant that most cases proceed to a final determination.

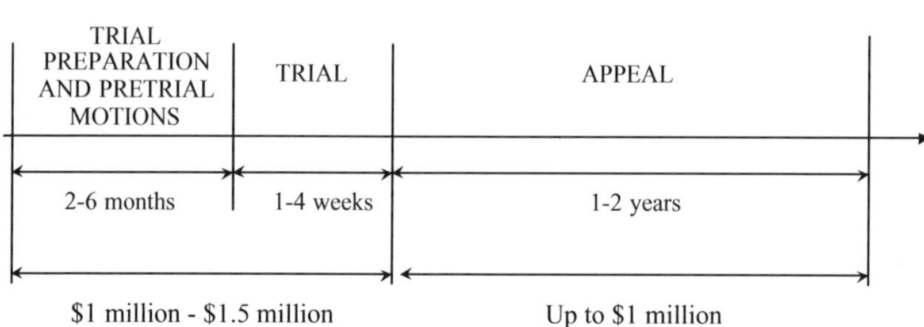

Figure 2.2

not fall within the claims. Questions of patent interpretation, validity, and damages may spur tenacious dispute, of course, but their resolution does not require extensive intrusion into business operations. Such a case may be brought and successfully litigated in virtually any country where a patent is in force.

Not so if the patent covers something locked away in the defendant's premises or a process carried out behind closed doors. In such cases, infringement may not be apparent from inspection of what the defendant sells, since products often do not bear traces of the process that produced them.

If some form of discovery is important in these patent situations, it is often indispensable when enforcing trade secrets. A trade secret *cannot* leave a detectable signature on products that result from its use, otherwise it would not remain secret very long. It will be invisibly employed behind the scenes or irreversibly buried deep inside a product. Detecting its unauthorized use requires the kinds of prying in which discovery specializes. The errant e-mail, revealing memoranda, the nervous witness rambling on one sentence too long – small victories such as these may not settle the ultimate issues, but often point the way to further information and witnesses until a clearer picture of the defendant's activities emerges.

One conclusion we can draw, then, before even reaching the strategic advantages and disadvantages of the various protection options is this: the degree of discovery available where IP rights will be enforced generally does *not* influence a rational decision between patents and trade secrets. The kinds of activities and hidden product attributes that make the choice possible at all resist detection under either IP regime, and, in countries that provide limited or no pre-trial discovery, evidence of their use may simply be unobtainable.

At the same time, litigation risk extends beyond the availability of discovery. The risks attending enforcement of patents and trade secrets do differ – enough so, very often, to bear strongly on the choice of protection mechanism. More on that below.

Patents v. trade secrets

CASE STUDY #1: Dmitri Ovkorsky, founder of True Blue, Inc., imagines a future in which we carry all the information dear to us – credit cards, driver's license, our music collections, maybe a movie or two – on a coin-size CD along with the change in our pockets. The key to this vision is a commercially viable blue laser, whose short wavelength allows it to read and write vastly more information on an optical disk than today's technology (which is based on red lasers). The semiconductor industry's quest for this grail-like goal has foundered on the fussiness of gallium nitride, the blue-emitting material of choice. Unlike silicon and other common semiconductors, gallium nitride resists efforts to form it into large, wafer-size crystals with high purity and few defects – characteristics necessary for lasers that actually work. While most semiconductor wafers are built up by depositing layers of the material onto a platform, like silt caking up on a lake bed, Ovkorsky takes a different approach, slamming precursor reactants together at high pressures and temperatures until blobs of gallium nitride form and grow, like snowballs. These can be cut up into thin, crystal-clear wafers of pure gallium nitride.

While not the first to form crystals in this fashion, Ovkorsky's many years in Russian military research exploring precisely this problem put him far ahead of commercial researchers just beginning to try the bulk-growth approach. He believes they are years from duplicating his "recipe," which involves subjecting the reactants to very specific pressures and temperatures in successive cycles. But these cycles are not fixed; instead, they vary depending on how well the crystal is growing. Some very fancy neural-network software, devised by Ovkorsky, continuously monitors some very fancy sensors in the reaction chamber to maintain precisely the right conditions.

The marketplace is enormous. Global sales of conventional red lasers presently exceed a billion units per year, and when supplanted by the demand for blue lasers in a still-growing market, the potential for True Blue's technology is astronomical. True Blue is still a young company, having been founded by Ovkorsky almost immediately upon his immigration to the United States 18 months ago. Money was not hard to come by and True Blue has enough to last about a year (at its current "burn rate").

You, the recently installed CEO of True Blue, are responsible for crafting and executing the company's IP strategy. Next week the board of directors expects your thoughts on the most basic threshold question – by what means will True Blue protect its proprietary technology?

You realize right away that the decision is not monolithic, since True Blue has more than one distinct technology – its process recipes and its control software. While obviously interrelated, each presents its own IP considerations.

But first some basic eligibility criteria come to mind. The Russian Federation, you recall, requires a foreign-filing license before patents can be filed outside Russia. Where, exactly, was the technology developed? Better dig into that.

Ovkorsky has been in the US for 18 months – a long time in the world of research and money. Did he publish anything? What about investor presentations? How much did he tell people, and under what circumstances? Has True Blue undertaken any activity that could remotely be characterized as a "sale," nascent as the technology may be? All of these factors will bear on the availability of patent protection, regardless of its strategic merits.

Let us assume these "housekeeping" details do not preclude eligibility for patent protection. Now we can explore the factors favoring patents and trade secrets, and see how they apply to True Blue's situation.

By now, of course, as CEO, you have an excellent feel for the competition. They're big players – Sumitomo Electric, Sony, Mitsubishi, General Electric – and universities, which is no wonder, given the stakes and the expensive equipment needed for research and production. It worries you, of course. You imagine the equipment Sumitomo's Advanced Material division must possess, all of the enormous reactors, warehouses of exotic gases in tanks, clean rooms with people milling about in white hazard suits. But of course you can only imagine, since Sumitomo isn't about to give you a tour. Which worries you further, for a different reason.

"Dmitri," you ask, "can we tell if someone is stealing our secrets?"

"Of course," replies Ovkorsky. "If they make crystals as nice as ours, they must be using our process."

"Can you be so sure? I mean, does the process leave some kind of tell-tale trace on the crystal, so you can figure out how it was made?"

Dmitri fixes you in a hard gaze. "Only our process can form perfect crystals. There is no other."

"Today."

"Yes, today. Who can predict tomorrow?"

"How long is it before tomorrow?" you ask. "I mean, how long before our competitors either catch up or figure out a different way create what we create?"

Now Dmitri is thoughtful. "Of course, this is hard to say. I like to think never. That's naïve, of course. Sometimes the mere fact someone else has solved a problem leads you to the same solution, independent of his efforts. Someday, probably, someone will duplicate our recipes."

"What about our software?" you ask, but it's too late. Dmitri is stalking off to the lab.

This brief conversation gives you a lot to think about. If you patent the recipes, everyone will know them. Your IP lawyer told you so. You thought you were clever when you asked whether you could combine patents and trade secrets, omitting the "secret sauce" of the recipes from the patent application and just giving some general guidance. But no – especially in the United States. A patent must teach the *entire* invention – all of it, not just parts. And not merely ways someone might hypothetically practice the invention; US law demands disclosure of the *best* way Dmitri knows of, the optimal recipes he has discovered, on the day he files the application.[3]

The penalty for holding back is draconian: in the United States a patent that fails to disclose the best mode is invalid *in its entirety*; every claim – not just the ones directly related to what has been withheld. Of course, the patent office will probably not learn of the omission during prosecution of the application. But, as soon as an attempt is made to enforce the patent, depositions and document requests will almost certainly reveal the basic concealment. Patents and trade secrets can rarely co-exist; it is one or the other.[4]

The best-mode requirement may seem especially frightening when looking ahead 18 months to the all-but-inevitable publication of the patent application, which all-but-inevitably occurs well before the patent office has reviewed it – so it is impossible to know whether the protection it will offer is commensurate with the subject matter disclosed. This particular fear can be managed, however, by doing what the patent examiner will do before she does it: perform a literature search to determine whether the invention is, in fact, an invention at all. As it happens, True Blue's patent attorney performed just such a search and gave you the encouraging results for a (relatively) modest fee. While no search is perfect, you were told, it provides a solid basis for weighing the risks and rewards of the patent process.

So patent protection is probably available, but trade secrets are an option as well. And it is either, but not both. So far we have not made much progress.

As noted earlier, the availability of a legal right is not the same as the ability to uphold it. Unless True Blue can detect outsiders' use of its process recipes, a patent covering them is not just worthless; it is suicidal for the business, a free instruction manual broadcast to the world with no mechanism for enforcing the exclusivity it provides. True Blue's process does not leave fingerprints. No CEO will ever peek inside his competitors' reactor rooms. Formal litigation discovery might be the only way to pry the doors open. But what about the courthouse doors? If they remain closed, which they will unless you can

[3] Note that the US "best mode" requirement applies regardless of the applicant's home country. So, even if the applicant hails from a more permissive jurisdiction, his decision to file in the United States subjects him to the best-mode requirement; his US application may therefore be more extensive than the one he filed in his own country.

[4] Of course, a patent may be compatible with *later-developed* trade secrets, since the teaching obligation ends with the filing of the application.

develop a good-faith basis for suspicion based on publicly available information (or a fortuitous tip from a disgruntled employee), you will have no basis to sue.

So far a trade-secret approach is sounding pretty good. True Blue can take steps to keep its manufacturing secrets under wraps. Competitors may conceivably come up with Dmitri's recipes on their own, but strict security precautions and ironclad employment agreements will prevent or at least impede their leaking out of True Blue. That is not true in other businesses. Pharmaceutical companies are not free to keep the compositions of their drugs secret; government regulators, not to mention prescribing physicians, insist on knowing. Were the maker of a new telecommunication switch or router to decline to reveal its design, no one would buy it; telecom customers must be convinced that the device not only works as advertised, but will conform to standards and play nicely with existing equipment. Some enterprises, in other words, have a business need to disclose their secrets.

The picture starting to form in your head looks like this:

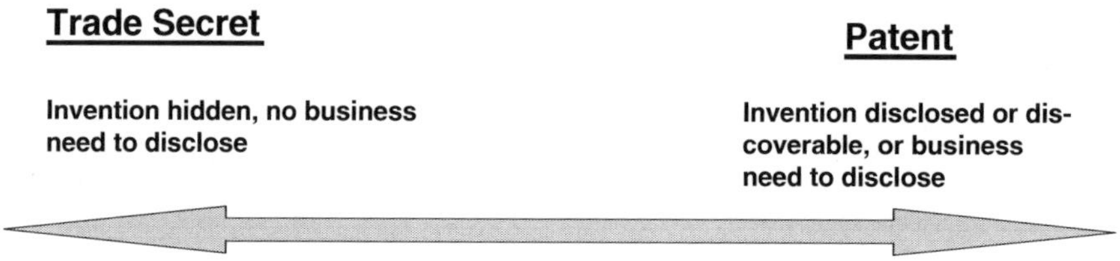

Trade Secret

Invention hidden, no business need to disclose

Patent

Invention disclosed or discoverable, or business need to disclose

On the other hand, maybe that right-hand arrow should not be dismissed out of hand. Perhaps Dmitri was not merely boasting when he said there is no other way to manufacture perfect crystals. If, indeed, perfection could itself serve as evidence of theft, then maybe patent protection is viable after all. You could walk into court with a competitor's crystal and argue that quality alone incriminates its maker.

Moreover, the ability to police a patent means not only detecting infringement, but being able to identify a relatively small number of deep-pocketed infringers that dominate the market. That is exactly the type of niche in which True Blue operates. Once again, many other businesses do not have this advantage. What if you had a patent on something anyone could throw together in his garage? Or a patent on software that hackers could duplicate and give away to your customers? Competing producers make highly unattractive litigation targets unless their sales – which dictate the damages you can recover – outweigh the high cost of a lawsuit. End users cannot be sued as a practical matter unless the community is extremely small, and, in any case, suing actual or potential customers is never good for business. Without a market structure conducive to enforcement, therefore, a patent can all too easily serve competitors better than its owner. (A trade secret, while also not cheap to enforce in court, at least is not teaching anyone anything.) For True Blue, the market is a factor favoring patents.

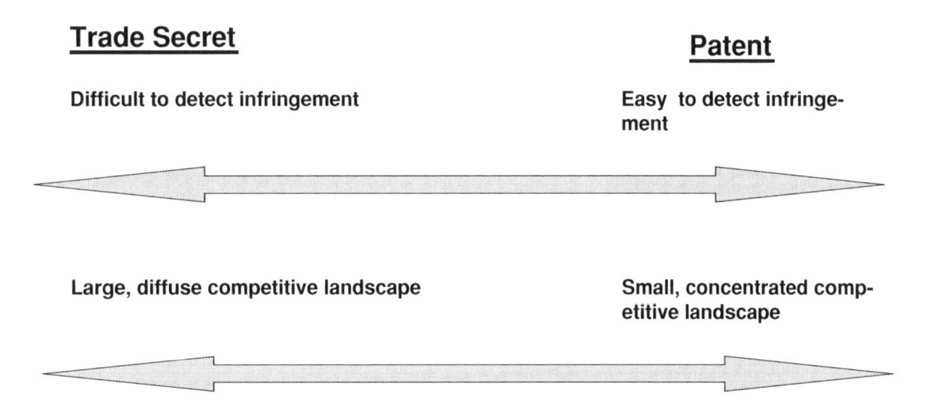

Still another factor in favor of patenting the recipe is the prospect of licensing. True Blue may have grand ambitions, but surely it does not imagine supplying the entire world demand – more than a billion lasers a year – by itself. If True Blue wishes to see its solution adopted in the marketplace, it must of necessity assure the marketplace of a reliable, redundant supply. The need to license technology to other companies, even its competitors, seems a pragmatic inevitability. A patent is licensed more easily and with less risk than a trade secret: more easily because exclusivity is legally ordained and not subject to sudden loss through disclosure, and with less risk because third-party licensees need not be brought into a circle of trust.

"We didn't get a chance to talk about our software," you remind Dmitri.

He shrugs. "What's to talk about? It is brilliant, yes, we cannot produce crystals without it. But . . ."

"But?"

"Well, it could run more smoothly. We're refining the monitoring algorithms. And of course our neural-network approach is only one of many that could be used."

"Wait. I thought we were the first to do this. Are you telling me it's all old hat?"

"No, no. Our implementation is totally new and original. But it's a work-in-progress. The general approach has been known for years, and, in any case, there are many ways to control a process. I almost lost a finger today because the pressure went critical and gas blasted unexpectedly out a relief valve." He sighs. "Maybe it really is not so brilliant. But it works most of the time."

"And others don't have it. Look, let me ask you this. If someone figured out our process recipe, how long would it take them to design your software? Or something equivalent?"

"This depends on how hard they work and how smart they are. A year. Maybe two."

"Take care of your finger," you tell Dmitri as you hurry back to your office. The board meeting is tomorrow.

True Blue's software clearly falls in a different category from its process recipe. When you first talked to your patent lawyer about IP protection, she thought you'd want to patent the software but keep the recipe as a trade secret. No one patents semiconductor process recipes, she told you, because they are usually just tweaks and optimizations – too trivial to patent and too easy to rip off if known. Yet, maybe it makes more sense to take the contrarian approach and patent the recipe rather than the software. A recipe patent could be policed, while the neural-network software represents merely one approach among many possibilities.

So many variables. So many factors bearing on the decision. This way or that way?

One thing you know about patents is that they take time: 3 years or so from the beginning to the end of the process. Clearly patents make sense only for technology having an expected lifespan of at least a few years,[5] and probably more than that – after all, a patentee must be able to earn a return on its investment. The focus, however, should be on the lifespan of the enabling technology rather than that of a specific product. Key innovations underlying performance improvements or capability enhancements often persist through multiple short-lived product generations, so that basing the patent decision on any single product in isolation will shortchange the long-term strategic benefit of patent protection.

From Dmitri's comments, it sounds as if True Blue's control software continues to evolve, and may be completely superseded at the most fundamantal

[5] One qualification to this tenet is the availability, in some countries other than the United States, of "petty patents" that issue more quickly and cheaply than conventional patents. See chapter 4. Only in rare cases, however, will this affect the overall calculus between patents and trade secrets.

level by the time a patent issues. Besides, it is not as if Dmitri has found the one and only way to control crystal production. Think about it. You get a patent and successfully defend the territory it covers. No one else can use True Blue's approach. So they will simply develop an alternative. Unless a patent covers the only commercially feasible way of doing something – whether because it is the one way that works or because other approaches will cost more, or take too much time, etc. – the patent has not accomplished much. There is little reason to go to the expense of a patent if it will not keep competitors out of the game, or at least at a market disadvantage.

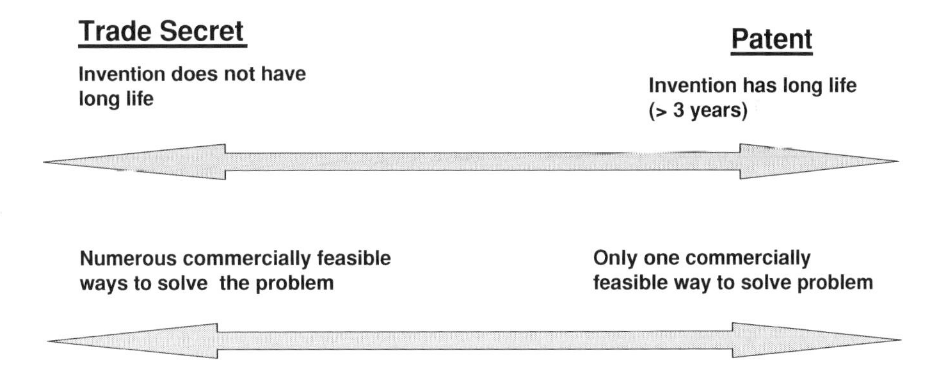

Finally, although you hope you will not be called upon to litigate True Blue's IP, you know you should consider the impact a lawsuit will have on IP assets. A patent owner asserting infringement is daring his opponent to challenge the patent as invalid or unenforceable. And, once killed, it cannot be revived. A patent declared unenforceable is useless against anyone; the victorious defendant is far from the only winner. Of course, the more literature searching that is performed before filing the application and the more thorough the patent disclosure, the lower this risk will be. A trade secret owner, on the other hand, must be prepared to define and disclose the details of the trade secret to the court and the defendant from the outset – generally under conditions that will prevent it from entering the public eye, but no system is perfect. In addition, a trade secret owner (unlike a patent owner) must vigilantly pursue any misappropriation of rights; failure to do so may result in a waiver of trade secret status altogether.

These risks strike you as interesting but somewhat theoretical, at least for True Blue, and do not bear too heavily on the decision between patents and trade secrets. The choice seems clear.

"Glad you finally made up your mind," says your IP lawyer. "It sounds like you've considered all the relevant factors. Patents for the recipe, trade secrecy for the software it shall be."

"Thank you."

"I admire a person who isn't afraid of forfeiture."

Forfeiture. What's she talking about?

"Oh, sorry," she says after an awkward silence. "Forfeiture is what happens when a trade-secret owner is patented out of business by a newcomer."

"I don't follow," you say after more silence. "How can a newcomer stop us from doing what we've already been doing?"

"Crazy world, eh? But patent law so favors disclosure that a later patent applicant trumps an earlier secret user. The patent applicant wins because of his willingness to share what he knows with the public – at the expense of the secret user."

"Well, what if we apply for a patent, too?"

"At that point you can't. Your prior use of the software for profit, even if it's all done in private, bars you from getting a patent."

"Wait. So my activities bar me, but not the newcomer?"

"I told you it's a crazy world."

Well, that's a fly in the trade-secret ointment. It would be no fun to be told you can no longer use your own software. You do a little reading and find that the threat of forfeiture underlies the strategy of "defensive publication": companies often pay obscure journals a nominal fee to publish the details of ideas currently too speculative or uneconomical to patent, but which might be useful some day. Early publication precludes *anyone* from obtaining a patent on an invention – forever. But disclosure, of course, is the antithesis of trade secrecy.

You also learn that US patent law also now contains an exception to forfeiture for patented "business methods." While the scope of this exception is not presently clear (the statute not-so-helpfully defines an exempted business method as "a method of doing or conducting business"), it almost certainly does not cover True Blue's industrial-grade activities.

Thinking long and hard about the dread prospect of forfeiture, you decide maybe it is a worry you can live with. After all, every trade secret is vulnerable, and people have not stopped keeping trade secrets. Moreover, a competitor deciding between patents and trade secrets would be faced with the same trade-offs as you, and probably would end up in the same place – that is, if you decide to protect the software as a trade secret, your competitor (hopefully) will, too.

Take a deep breath. The board meeting is tomorrow and we still have not considered copyright.

Patents v. copyright

Comparing patents to copyrights generally means software is involved. Only rarely will copyright overlap with other forms of patentable subject matter. True Blue's recipes, for example, benefit little from copyright. Write down the recipe conditions and procedures and the writing will be protected from duplication, but the content will not be guarded against execution. True Blue's control software, on the other hand, may be fair game for both the copyright and patent regimes.

Stretching the corners of copyright to cover software has never produced an especially comfortable fit. Although computer programs can be perceived and comprehended by humans, their chief function is not to inspire human thought, but to direct the operation of digital machines. This utilitarian role runs counter to the spirit of a system intended to protect expressive creations. Copyright was never meant to cover subject matter that "does" something; that is what patents are for. The patent system offers only a limited period of protection, and sets rigorous standards of invention that must be satisfied before exclusive rights are accorded. The duration of copyright, in contrast, is long (and getting longer), and the qualifying standard is minimal.

Copyright has long dealt with subject matter having both informational and functional aspects, covering the information but not the function. Architectural blueprints, for example, fall within copyright, but not the structures they portray. An architect can prevent unauthorized duplication of plans for a house, but cannot prevent a legitimate owner from actually building it. Form is separable from function because a blueprint's information content is distinct from the building it portrays. Not so with software. Rather than merely describing the operation of a machine, software is more like the machine itself.

Still, software is a written work subject to easy duplication and dissemination – just like a poem or a book. In this sense, the copyright paradigm is entirely apt: copyright law is, of course, all about copying and its unauthorized exploitation.

Placing software under the rubric of copyright, therefore, subverts the law by effectively extending it to functioning computer hardware; one cannot separate the functionality of software (what it does) from its expressive content (what it is). Excluding software from copyright, however, would ignore its textual attributes and the ease with which it may be copied.

The decision, of course, has been made in favor of software copyright – both for software as text (that is, program instructions and data) and for the screen displays that it causes. But the underlying tension remains, and manifests itself in the degree of protection actually afforded – which, basically, is not very much. To understand why, consider first how a novel or play is protected. If Shakespeare were alive to copyright his works, his quill would not foreclose others from expounding on the death of Julius Caesar, nor would *West Side Story* infringe *Romeo and Juliet.* But were someone to copy more than the basic structure of the plot – the idea – and appropriate some of the recognizable quotations and expressions, copyright would spring into action, and for sound intuitive reasons: why should someone be able to profit from the gems Shakespeare put into Mark Antony's mouth? Go write your own soliloquy.

The richness of written and spoken language guarantees that many other versions of Mark Antony's speech could be imagined without loss of fidelity to the basic plot. It is for this reason that duplicating an author's particular choice of words strikes us as wrongful. Computer programs, however, aim not for soaring oratory but to get a job done – to control registers and push bits and generate output. There may be very few ways of doing that job. Were copyright to prevent even slavish copying in such cases, it would be protecting the machine and not the expression. (In fact, there *is* no expression here; it is merged with the idea, that is, the operations being performed.)

Consequently, when analyzing copyright claims involving computer software, courts put the program through a preliminary sieve-like analysis to determine which elements qualify for copyright protection. Portions that cannot be coded in more than one or a few ways are discarded. So are common, basic routines to perform well-characterized tasks, such as storing and retrieving data. The same goes for routines whose form and content are dictated by outside factors, such as industry practices or standards, or hardware requirements. And that is just at the code level. High-level concepts, such as algorithms, mathematical equations, and formulas – the heart of the way technical problems are solved – represent pure function (rather than expression) and lie outside copyright altogether. What is left is a middle ground, kernels of creative expression that the court compares with the accused program. A finding of "substantial similarity" to these elements along with access on the part of the accused author will result in an infringement verdict. The bottom line is that copyright protects implementation far better than function. The bigger the implementation and the more slavish the copying, the greater will be the chance that protectible elements will be found to have been infringed. Duplicate a Microsoft Office CD-ROM and copyright liability is guaranteed. Create a new control program to operate True Blue's production equipment and things are far more uncertain – especially if the program is developed without access to True Blue's software.

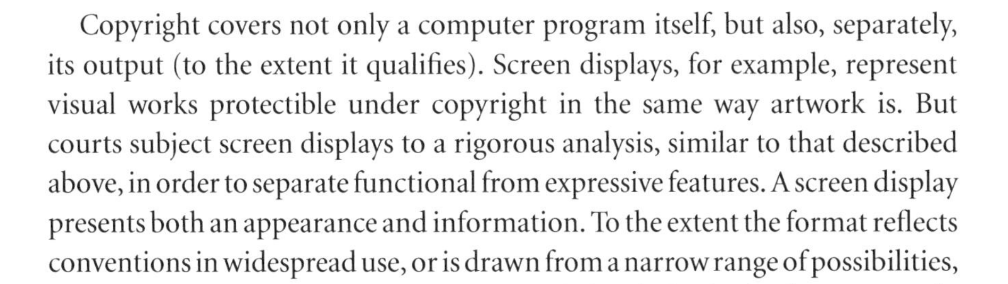

Copyright covers not only a computer program itself, but also, separately, its output (to the extent it qualifies). Screen displays, for example, represent visual works protectible under copyright in the same way artwork is. But courts subject screen displays to a rigorous analysis, similar to that described above, in order to separate functional from expressive features. A screen display presents both an appearance and information. To the extent the format reflects conventions in widespread use, or is drawn from a narrow range of possibilities, it will not receive significant protection; while wholesale duplication might subject the duplicator to liability in such cases, even minor changes will allow him to avoid infringement. Arbitrary, artistic content receives commensurately broader protection.

An important twist in Europe, which affects both computer code and screen displays, involves interoperability. European copyright law expressly prohibits restrictions that would inhibit reverse engineering to achieve interoperability with other programs and hardware. That means copyright law will not help software developers seeking to construct proprietary walls around their products – interfaces, for example, erected to prevent rather than facilitate competitors' efforts to create complementary or compatible products.

Copyright protection for software, then, is relatively limited in scope. But do not assume that what stands outside copyright necessarily falls within the patent system. Although patents do largely pick up where copyright leaves off, some subject matter remains off-limits even under the patent system:

- *Mathematics and laws of nature.* You cannot patent algebra and you cannot patent gravity, but you can certainly patent applications of either – a ballistic missile, say, equipped with trajectory-adjusting software and which lands thanks to the pull of Mother Earth. Similarly, True Blue's process may be based on the action of physical laws, but it is the resulting crystal growth and not the laws themselves that we would patent. While copyright protects expression but not function, patents protect function but not underlying principle.
- *Business methods.* Software and systems implementing e-commerce, marketing techniques, financial analysis, and pretty much everything else are fair game for patenting in the United States. Elsewhere, and particularly in Europe, patentable subject matter must be "technical" in nature. While the

line between technical and non-technical subject matter is difficult to draw and often somewhat arbitrary, some examples may serve as guideposts. Software for helping a bank's customers choose among financial instruments would be too business-oriented to patent outside the United States. A spell-check program probably still would not be technical enough for Europe, likely would pass muster in Japan, and would certainly qualify in the United States. True Blue's software is most assuredly technical in nature, controlling as it does a complex industrial process, and would be patentable anywhere.

- *Software code.* The United States has firmly established the patentability of program code and Europe has roundly rejected it. The law is less clear elsewhere, but it is rarely necessary to patent code in order to protect software. Rather, protection focused at the machine or system level – that is, what the software makes a computer do, how True Blue's software controls their process – is usually adequate.

Does relying on copyright protection make sense for True Blue? Probably not; we already know that True Blue can achieve similar results in different ways, meaning that even ironclad copyright protection would do little to deter others from doing the same thing using different code – indeed, using a different approach altogether. Conversely, if True Blue's way were the only way, it would lie outside copyright law altogether (that merger problem). If you were inclined toward trade secrets, nothing in the world of copyright would change your mind.

Copyright v. trade secrets

This is easy. There is no conflict; you can have it all. Copyright applies to published and unpublished works alike, and software distributed in a manner that preserves trade secrets – for example, with a license prohibiting decompilation and reverse engineering – enjoys full copyright protection as would, for example, an unpublished diary. Copyright, in sum, is compatible with trade secrecy. There may be some sorting out to do when a lawsuit is filed (where does the copyright claim end and the trade-secret claim begin?), but the systems do not contradict each other.

In fact, the systems collide at all only in connection with "deposits." Recall that in the United States it is necessary to deposit a copy of a work with the US Copyright Office in order to obtain a copyright registration, which is itself prerequisite to an infringement lawsuit. How can you deposit what you do not want the world to see? The answer is that the requirements are flexible enough to permit retention of trade secrets. The Copyright Office welcomes submission of the complete program source code,[6] will be satisfied with the first and last 25 pages of source code, and will grudgingly accept the first and last 25 pages of object code. While some may be tempted to fill those first and last 25 pages with unimportant content or complete irrelevancies, the result may be a defective registration. Moreover, one function of the deposit is to establish ownership, a benefit to the registrant that fails to the extent that what is deposited deviates from the actual commercial product.

Not that any of this has much to do with True Blue, which has no plans to put competitors into business by selling its software. But should True Blue choose to protect its software as a trade secret, copyright will come along for the ride.

Here are the basic preference factors summarized, indicating where True Blue's technologies fall along each scale:

Process Recipes

Trade Secret	Patent
Numerous commercially feasible ways to solve the problem	Only commercially feasible way to solve the problem
Invention does not have long life	Invention has long life (> 3 years)
Invention hidden, no business need to disclose	Invention exposed or discoverable, or business need to disclose
Difficult to detect infringement	Easy to detect infringement
Large, diffuse competitive landscape	Small, concentrated competitive landscape
No need to license technology to others	Licensing to third parties essential

[6] For those unfamiliar with the distinction, source code represents instructions in English-like words and syntax, and can be understood by humans; object code is the result of translating source code into a long string of binary digits that actually operates the computer.

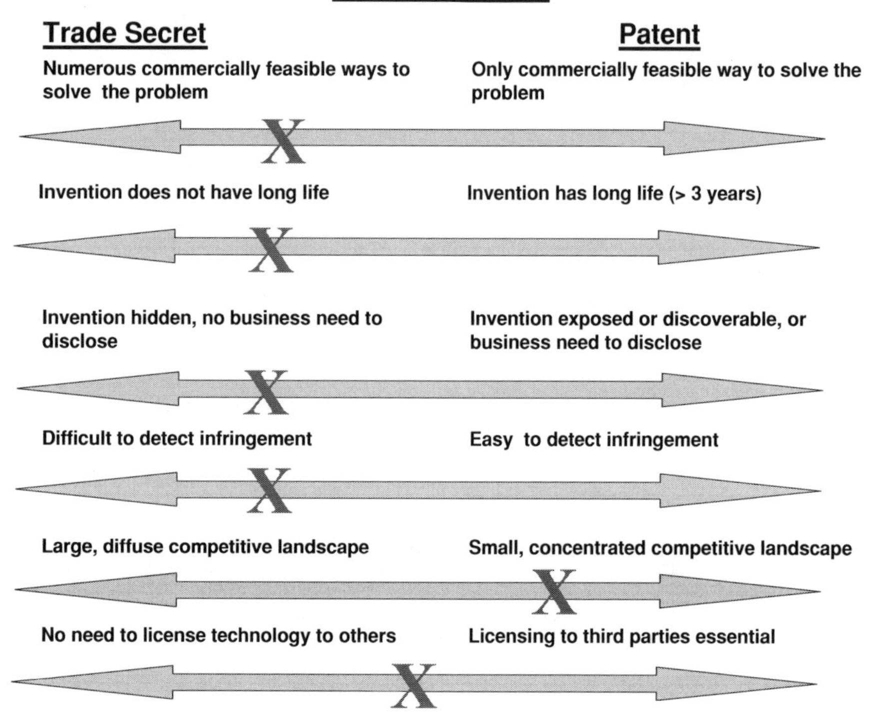

Control Software

Copyright | **Patent**

Large program subject to wholesale duplication | Value is in specific program functions, algorithms

X

With the trade offs clarified and set against True Blue's technology, the right choices emerge clearly. Patent the process recipes, keep the control software secret, and accept the limited relevance of copyright. Unruly alternatives have coalesced into a sensible plan. Winning the board over should be a piece of cake.

3 Building an IP strategy

A technology company must have its patent strategy, to paraphrase H. L. Mencken, as a dog must have his fleas. (Mencken was speaking of professors and their theories, but any theorist worth his fleas will doubtless sport a few patents as well.) The truth, however, is that few companies develop IP strategies that are business driven and responsive to changing circumstances; most, instead, indulge in guesswork and fall victim to inertia.

The consequences of pursuing IP without a strategy are predictable enough. Patents make expensive wallpaper. A portfolio that grows haphazardly will ultimately diverge from what is important to business success, covering the past rather than the future and inflicting no pain on competitors. What appears formidable in terms of numbers of patents or their cost may be no more than strategically useless decoration.

How to achieve this unhappy state of affairs? One excellent way is to view IP procurement as a scientific enterprise rather than a business function. Top managers may be tempted to cede full responsibility for IP to the chief technology officer or, worse, to fragmented groups of engineers, each operating with its own budget and discretion. The approach seems superficially sensible, since those others know about patents and technology, and lord knows the managers already have their hands full. But scientists often pursue the intriguing rather than what is important – that latest technological twist may engender appreciative oohs and ahs but no business benefits realizable through IP. For example, as noted in the previous chapter, IP covering the most ingenious solution will not be valuable if other, less ingenious but commercially serviceable solutions remain available. Competitors will merely sidestep the IP barrier and adopt an alternative.

Managers who treat IP as something for the techies to worry about – who, in effect, see IP as a commodity or cost of doing business – are inviting stiffer-than-expected competition, patents that lack business relevance, and difficulty attracting licensees and/or business partners. Managers who recognize the dangers of an aimless IP program but nonetheless refrain from direct involvement often turn to consultants bearing colorful PowerPoint slides and prepackaged strategies. Without the very participation such managers hope to avoid, however, these "strategies" will be no more than empty suits. Still, they have attained such visibility, it is worth reviewing some of the more fashionable styles and what they purport to fit.

The "core technology" strategy advocates identification of what the company does best – its "crown jewels" – and intense efforts to obtain foundational patents in these areas; in other words, if you make computer chips, patent everything you can about the chips. The "target" strategy surrounds the core with a series of circles corresponding to zones of diminishing priorities. Following a review of the company's product lines and current research directions, candidate areas for IP exploitation are assigned to one of the priority circles. The first ring around the bull's-eye core may encompass implementations of the foundational technology or features that optimize or facilitate its exploitation in terms of performance, ease of use, cost, or some other measure. Another circle may represent potential applications, another incremental improvements, and so on.

The "picket fence" approach means different things to different people. To the cautious field marshal, it means filing families of patents around a key product instead of asking too much from a single patent. To the swashbuckler, it means taking the fight to the enemy, filing patents around *competitors'* core technologies. And, as a strategic retreat, it can mean erecting obstacles around a weak core technology; if you can't patent the chip, in other words, patent the interfaces and communication buses that facilitate interaction with it. In this incarnation the picket fence looks suspiciously like one of those target rings around the bull's eye.

These strategic chestnuts have the reassuring appeal of a coloring-book outline – just fill in the specifics and watch a gorgeous picture emerge. Unfortunately, like such outlines, they portray the obvious in a simplistic fashion and straightjacket the ability to think outside the borders they inscribe. If you can not protect your key technology, *of course* you are going to protect what you can in an effort to ward off competitors. And *of course* you are going to prioritize among projects. All of these approaches have validity and all have their place. But when you partition the world so it looks like a picture, rather than examining whether the picture validly portrays the world, you are asking for trouble. For example, deciding what is central and what is peripheral can be a very tricky business. "Core" technology usually means everything that is or will be essential to fulfillment of the business plan – innovations embodied in current products or those on the drawing board. But, ultimately, separating the essential from the ancillary represents no more than guesswork. Many

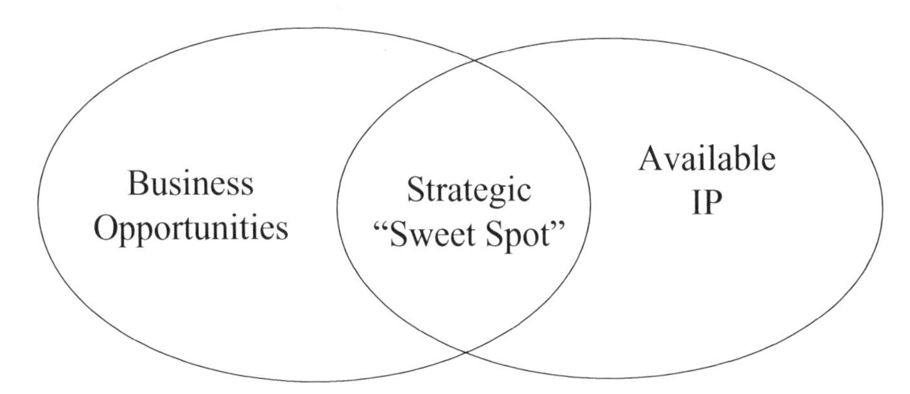

Figure 3.1 Aligning IP opportunities with business opportunities

star patents were born in obscurity, mere "throw-ins" that looked inconsequential at the time they were filed. Much also depends on who is ordering the rankings and their pet priorities – business v. technical merit, short-term v. long-term value, competitor v. internal focus. And, while any strategy can fail the freshness test if it is not constantly reviewed, the more simplistic the approach and the more seductive the metaphor (build me a leakproof wall!), the more readily can it turn into a mechanical exercise – the mind becoming a slave to the metaphor and strategy a servant to the picture, which is not rethought, regardless of how stale it becomes.

A grand strategy for developing a strong, balanced IP portfolio amounts to no more than this:

IP protection must fully and constantly align with the company's business objectives.

A very simple proposition in words or pictures (see figure 3.1), though not, as we shall see, so easy to implement. What this means is that IP opportunities always exist outside what is sensible to pursue, and not all business opportunities will correspond to IP opportunities. Identifying the fertile grounds requires understanding the strategic benefits that IP can deliver as well as a detailed picture of a business's operations and marketplace objectives.

What can IP do for you?

In the rush to establish IP rights, the reasons for doing so can be overlooked. IP is never a panacea, nor are the roles it can play in the life of a business always self-evident. Considered at a high level, those roles can be both offensive and defensive. First the offense:

- *Wall out competitors.* This is what most people hope for when seeking IP protection. Trade secrets provide at least a head start against competitors, patents a fixed but usually adequate period of exclusivity, and copyright is nearly forever. "Blocking" IP that clogs a bottleneck – covering something would-be competitors need in order to compete – provides the tallest wall, and careful procurement practices make it strong.
- *Install a tollbooth in the wall.* It is fun to imagine deriving licensing income from noncore technologies – that is, IP you were not going to exploit anyway. Such windfalls are unusual. Circumstances may, however, warrant licensing even the crown jewels to foreign manufacturers or distributors, development partners in specialty markets, anyone with access to an audience you cannot reach – or even to all comers, depending on the size of the market and barriers to entry.

Now the defense:

- *Intimidation.* A company with a patent portfolio makes a less inviting litigation target than a "naked" competitor. The possibility of a countersuit raises the stakes and may ward off an aggressor entirely, or at least serve as leverage in settlement negotiations.
- *Access to others' technology.* If you have valuable IP, you have something to trade.
- *Stop others from stopping you.* If you have something to trade, maybe you can cut a deal with the owner of blocking IP. Moreover, an early patent filing date not only (hopefully) gets you worthwhile protection, but can protect you against patents subsequently filed by others. (Of course, simple defensive publication can achieve the latter objective.)

And also to be considered:

- *Dealmaking.* When negotiating technology partnerships and joint ventures, IP is what you bring to the table. Suppose you are hoping to partner with a reseller that will incorporate your company's technology into a commercial product. A strong patent portfolio not only provides the potential partner with the prospect of market exclusivity (or at least IP leverage against competitors), but also discourages them from partnering with *your* competitor or developing the technology themselves.
- *IP as an asset.* IP can serve as collateral, as a valuation basis for potential acquirers, and as psychological support to nervous investors who wonder whether there really is anything under the hood of your company.
- *Customers may insist.* Yes, big customers sometimes demand patent protection on what they buy, particularly if the procurement is large and the

integration cost high; they do not want to see their cost of doing business (the price of your company's product) undercut for their competitors by your competitors.

- *Politics.* IP should not be used to bestow recognition, but some inventors expect it.

Characterizing the business

The pulse of any new business is most easily taken through examination of its business plan. A competently written plan lays out the entire circulatory system of the marketplace and the company's place within it – the needs of customers, the channels through which goods or services materialize and reach them, competitive distinctions, the special skills and expertise that the company brings. For an established firm, this anatomy is already well characterized; start-ups must conjure the particulars as realistically as they can from prior business experience and industry research. Specificity is essential to an effective strategy because IP, too, is specific: patents, for example, cover products or techniques for making them, not the overall business. Appreciating the contribution of IP to the health and function of its owner, then, demands an X-ray rather than a wide-angle snapshot of the "value chain" of activities that turns inputs into value-added outputs for customers. For example:

raw materials → design → manufacture → distribution → customer → recycle/reuse, service

By considering business objectives on a segmented basis, the relevance of IP can be more firmly established by assigning it to one or more activities in the sequence. In this way its contribution may be put into proper perspective and assessed.

Putting it together

CASE STUDY #2: The Human Genome Project has been completed, laying bare the full genetic blueprint that underlies all of our traits and, in effect, defines personhood. Scientists can study the three billion chemical base pairs that make up human DNA – our genes – to determine how we develop, why we age, and the causes of disease. A tremendous achievement for humanity generally; less compelling for you in particular. Although 99.9% of the sequence is the same for everyone, that 0.1% of difference determines how *you* will react to particular medication, whether *you* will have a healthy baby or go bald or develop Alzheimer's Disease. Using the techniques of the Human Genome Project to sequence your

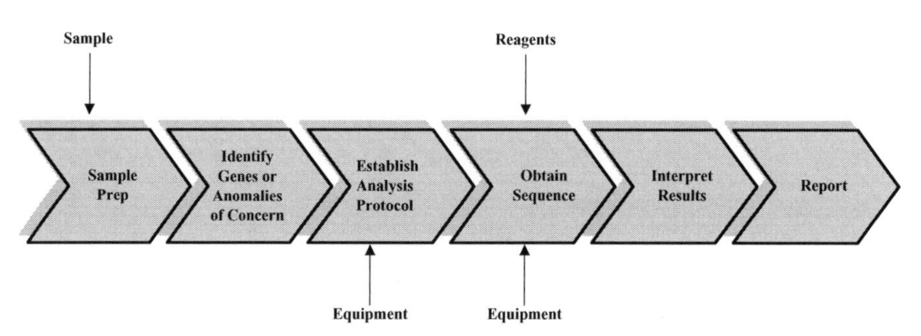

Figure 3.2 RSS's value chain

personal genome would require far more genetic material than you have to spare, and far more time than you'd like to wait (the project took 13 years).

Which is why Ridiculously Small Samples, Inc. (RSS) is so excited about single-molecule DNA sequencing. If an entire genetic sequence could be elucidated from the DNA in just one or a few cells, a simple cheek scraping or a drop of blood would suffice. If the sequencing technique were well designed, it could be completed in a few hours. Of course, the almost quixotic notion of analyzing any single molecule, much less one with the extraordinary data density of DNA, seems a direct challenge to Mother Nature – and she's a powerful adversary.

RSS, therefore, proposes to fight nature with nature, relying on the molecular assembly machinery used by living cells to duplicate their own DNA. Where conventional sequencing techniques pull DNA apart, RSS effectively peeks at the duplication process as new DNA is built. This approach allows even a single molecule to be decoded – and, most importantly, in separate pieces analyzed in parallel. Sequencing a full strand of DNA base by base would take months or years; RSS's divide-and-conquer strategy reduces that task from billions to thousands or fewer steps that may be accomplished in hours.

RSS is not the only company pursuing single-molecule dreams, but they think they're the first to approach the problem in this particular fashion. Their innovation ranges from the basic technique itself to the chemical reagents – the "juice" – that nudge duplication along in a detectable fashion to designs for detection equipment. Their business plan is still being written. What are the options, and how do they help shape an IP strategy?

RSS has numerous alternatives in terms of conducting business, where to focus its efforts, and with whom (if anyone) to partner. These options are clarified by making the value chain explicit (see figure 3.2).

Although someday everyone may want to walk home with his genome sequence on a CD, it is more likely that individuals instead will be tested for specific conditions – susceptibility to particular forms of cancer, likelihood that a particular treatment will succeed – corresponding to telltale genetic clues. A subject, let us say a physician's patient, provides a small tissue or blood sample. The physician identifies a condition of concern; maybe she also specifies the corresponding gene or genes, but more probably that is done by someone else – the laboratory performing the actual test, for example, or the sequencing equipment itself. The targeted condition determines how the sample is processed, confining the sequence analysis to the DNA regions of interest to save time and "juice." The raw sequence data are interpreted and the results provided to the physician.

The business possibilities for RSS are boundless. It can choose to do everything (receiving samples, analyzing them behind closed doors, and transmitting results) or nothing (licensing its technology to a manufacturer and passively receiving royalties). It can sell juice, equipment, or both into the marketplace. Perhaps the reagents are easy to make, requiring a small capital investment but providing, as indispensable consumables, a steady high-margin revenue stream – making them more attractive as manufacturing targets than equipment, whose production is likely to involve more capital and less volume. If RSS does not perform the actual analysis, will this task fall to independent laboratories in the manner of today's blood and tissue tests, or will every physician have a sequencer in her office?

So many prospects. We will not get into the mechanics of defining business objectives and developing a business plan, but we can explore how IP influences the critical decisions. Where along the value chain RSS focuses its IP program depends, of course, on the relative availability of protection at each link. Indeed, the 'sweet spot' in figure 3.1 – where IP availability intersects with business opportunity – may dictate not only technology-protection efforts but, often, business efforts as well. Interpreting a DNA sequence and reporting the results, for example, will likely represent mundane operations not amenable to IP protection. The market will quickly become unattractive as customers seek the cheapest provider of these commodity services. Moreover, even if RSS could obtain some meaningful coverage for its interpretation scheme, would it even want to? So long as it solidly protects earlier steps in the value chain – the core sequencing methodology, or perhaps even just the reagent formulations, assuming they are critical – RSS will likely opt to make its interpretation methods freely available in order to encourage widespread adoption of its technology. Ditto for sample preparation. These may represent IP opportunities, but provide little business potential.

What about gene identification? Forget it – thanks to the Human Genome Project, the relationships between genes and ailments will likely be public information, and a good thing that is for RSS. Those relationships underpin the basic usefulness of DNA sequencing. No one would purchase expensive machines and chemicals to perform guesswork.

The business opportunities of interest will center around the sequencing technique, the machines and juice to implement it, and protocols to operate the machines. These are the "core competencies," the ways in which RSS most beneficially distinguishes itself from its competition. Suppose that patent protection is available to RSS in all of these areas. Should it simply patent everything in sight? Possibly, but not without first thinking through the business objectives.

Let us say that RSS is hell-bent on world domination. Its goal is to manufacture all the machines and make all the reagents (although it will generously allow its customers to perform their own tests with the juice they buy). That means RSS wants a secure IP wall with no tollbooths. Viable strategy? If you are an RSS investor, those soaring aspirations might give you financial vertigo. Sure, if the basic development work is nearly complete and machines will not be too difficult to manufacture (probably offshore, under contract), RSS can construct pro forma financial statements that show year-to-year increases in sales and eventual market supremacy. But, in the real world, the strategy collapses under the weight of its own ambitions.

First, worldwide IP dominance is simply unrealistic. If you are a Big Pharma player, maybe you can spend millions of dollars to procure patent rights across the globe each time you develop a valuable product. Here we have assumed that RSS is likely to be able to protect multiple distinct aspects of its technology. Given the pioneering nature of its research – that's why all that patent protection tempts like low-hanging fruit – RSS will probably file several patent applications within each technology segment, translating into numerous patent applications, each to be filed in all geographic markets . . . the prospect boggles the mind and the wallet.

Enforcement is even more unrealistic. Every viable IP strategy must consider the possibility of disputes. The hotter the technology, the more likely its owner will be called upon to pursue infringers. A company's financial and psychic capacity for IP enforcement must match its market ambitions, lest it be overrun by competitors. Global enforcement is even less practical than achieving an enforceable worldwide IP position.

Finally, the fittest do not always survive. Time is the greatest enemy of all. In an ideal world, sheer quality would drive the best technology to the top of the market. In the real world, the marketplace moves on when opportunities are not exploited. Entice the market with the perfect solution but then fail to deliver, or don't deliver quickly enough, and the market will find another darling – even one that is less deserving. Corporate graveyards resound with the howls of "best of breed" technologies whose tardy owners succumbed to nimbler inferiors. An ill-timed lawsuit can have the same effect, discouraging customers from adopting a technology that's "in legal trouble." And even the hottest solutions cool rapidly in the breeze of a stampede. The larger the market, the more competition it will attract, and today's best of breed may be tomorrow's crowbait as well-heeled competitors engineer superior products. If RSS is determined to go it alone, it may become its own bottleneck; market entry and rapid production scale-up are notoriously difficult, particularly for new companies with scant experience. The mere delay inherent in an unrealistic strategy could be RSS's undoing.

A more realistic plan might be for RSS to produce and sell juice, based on the assumptions of high profit margins and small capital investment in line with its resources, but license use of the sequencing technique. Even this decision represents more direction than strategy, however, since there are so many ways to license. The traditional approach, certainly in the life sciences, is to partner with an industry giant and let them take care of product development, marketing, and crushing the competition in exchange for royalties and, sometimes, an equity investment. We will have more to say about the mechanics of such arrangements in chapter 7. For now the question at hand is what this business strategy, if adopted, would hold for IP strategy.

- *Perform a patent search.* Searches are entirely voluntary; no patent office requires them. They are usually a good idea, though, for the reasons discussed in chapter 2. And they are essential to a licensing strategy. To a licensee, your patents are not only what you bring to the party; they are its only line of defense against the onslaught of the market. A licensee is imprisoned by the quality of your patenting efforts and knows it. Not only knows it, but is quite certain it would do a far better job than you, or at least have no one else to blame for any shortcomings. Your job is to ease their skittish misgivings, convince them that you have taken every rational precaution to ensure patent strength. That begins with a high-quality patentability search, preferably performed by a specialist, followed by a thorough analysis from your attorney.
- *Fight for broad protection.* Of course you want broad IP protection on your core technology, whether or not you are licensing it. Who wouldn't? But remember that you are not creating a product for the marketplace; your licensee is. Which means that you do not really know what form the final saleable product will take. Neither does your licensee – yet. Part of the comfort factor you must provide is a solid sense that, when your patents issue, they will be extensive enough to cover any realistic implementation your licensee might choose. That means broad patent claims and a convincing patentability analysis supporting them.
- *Field-specific coverage.* In the world of licensing, IP can be sliced and diced to order: by geographic region, by product type, by market, and, perhaps most importantly, by field of use. That means RSS can license a patent (or

even a single claim of a patent) to one company for, say, cancer-screening applications and to another company for tests targeting heritable diseases and conditions. Indeed, RSS can segment the market still further, giving different companies the rights to develop, say, pre-natal and post-natal tests for a particular condition. The possibilities, in theory, are limitless. In practice, the number of licensees it makes sense to have depends on their comparative skills and advantages (it would be unfortunate to license *all* rights to a company good at some things but not others) as well as each prospective licensee's clout (which may make its demands for a broad field of use difficult to resist, even if greater diversity among licensees seems preferable). Although broad patent claims can be licensed selectively through contractual agreements, market segmentation is often easier to achieve if the patents themselves contain field-specific claims. This is because prospective licensees tend to think in terms of patents rather than technology. Field-specific patents reduce (to some extent) negotiations over how the field is to be defined, and may inspire greater confidence on the part of a licensee whose specific market segment is covered.[1]

- *Preserve international rights.* Perhaps RSS is short of cash (having diverted so much toward production facilities for reagents) and is tempted to confine patenting efforts to its home country or region. Big mistake when, for the price of a PCT application, it can preserve its foreign-filing options for the licensee. Once lost, such rights cannot be recovered, and their absence can dramatically affect the licensee's perception of value.
- *Identify third-party rights.* A license allows RSS's customer to use IP owned by RSS, but what about rights owned by others? A licensee wants complete freedom of action to serve the marketplace, not merely the assent of one stakeholder among many. Licensors should scour the literature for patents that complement or even overlap its own coverage – we will get into procedures for doing so later – and attempt to acquire them or at least partner with their owners to deliver a complete rights package to prospective licensees.
- *Coverage off the beaten track.* Although it seldom pays to devote too much time and IP effort to odd uses of technology or funky business practices

[1] Remember, each patent claim stands on its own. Suppose a broad claim is licensed to three licensees, A, B, and C, for different applications. Licensee A brings a lawsuit against an infringing competitor. The competitor, unfortunately, succeeds in invalidating the patent claim in the course of the lawsuit. In that case, licensees B and C are out of luck. Their patent protection has evaporated along with A's. If B and C had received licenses under different field-specific patents, on the other hand, their rights would not be subject to the vagaries of litigation involving another licensee. Indeed, a field-specific claim is by definition less susceptible to attack than a broad claim; recall that the narrower the claim, the smaller a target it presents to antagonists.

associated with it, a licensor can be less parsimonious in this regard than a commercial player. The licensor's "product" – what he has to sell – is his IP. Once engaged in an ongoing program of protection, a licensor's cost in extending coverage into unusual areas is often marginal.

Another approach to licensing involves contributing to industry standards. Any single licensee, however large, cannot control an entire market of any size. By making technology available to all through a standards organization or less formally, it is possible to reach the whole market. Assuming anyone cares, of course. The technology must be sufficiently exciting, and the price of a license sufficiently low, to entice a broad cross-section of the industry into adoption. Once a critical threshold is crossed, though, standards campaigns become self-reinforcing: what suppliers adopt, their customers (and makers of complementary products) grow to expect, forcing other suppliers to adopt as well. Standard-setting has become standard operating procedure in the intricately linked world of telephony and networking, where compatibility across products is a must for the entire industry; recognized standards organizations continue to grow and proliferate, regulating both approval and licensing practices. But similar market effects occur outside the tightly interdependent world of information technology. In life sciences, for example, techniques such as polymerase chain reaction, which amplifies the quantity of DNA in a sample – just the kind of process RSS is able to avoid – have become routine laboratory practice and earned fortunes for their originators. If the market is likely to welcome its technology even in the absence of a big partner, RSS will may wish to consider this approach. Of course, shouldering the burdens of an industrywide licensing program and, quite possibly, litigation can overtake every other aspect of RSS's business, consuming critical resources and distracting from further innovation.

Now let us turn to reagent sales. As a seller of consumables, RSS must think like a commercial player, not a passive licensor. The strategic mindset is different. To a vendor, IP represents a line of defense instead of a line of business, an expense rather than inventory. That means, with respect to IP covering reagents, RSS must strive for the greatest value at the lowest possible cost. Several basic principles guide this effort:

- *Cover your key markets.* As detailed further below, reagent-related IP should extend only to important markets. It is easy to overdo foreign protection.
- *Cover your commercial implementations.* Yes, you want broad protection. It would be nice to cover every possible design alternative. But do not discount the importance of IP coverage for what you actually sell. The hungriest competitors do not seek noninfringing alternatives, since workarounds require

actual work. Instead they copy. Strong patent protection for your commercial implementations will not only thwart those who nip at your heels, but protection will likely be relatively narrow – and therefore more easily defended. The broader a patent's coverage, remember, the more vulnerable it is to attack based on prior work.

- *Be alert for licensing opportunities.* A well-focused business plan excludes as much as it covers, if not more. The full measure of an idea's potential typically stretches well beyond the markets its originator can realistically serve. Others, however, may be in a position to exploit what even a commercially ambitious innovator cannot, and the possibility of licensing should always inform IP protection efforts. Still, a vendor is in the business of selling, not licensing, so the cost of protecting subject matter outside the sales cycle is not marginal. Every patent claim that does not cover what a seller actually sells is like a lottery ticket; the cost of that ticket is worthwhile only where markets are large and licensees can be readily identified. As an industry insider, a vendor may have better access to potential licensees than a passive licensor. The ideal licensing play for a seller is a line of business that would make it into the business plan if only the resources were available. At the same time, remember that failure to purchase the lottery ticket (or at least publish defensively) leaves it available for competitors, who may patent those peripheral applications themselves – walling them off. Suppose, for example, that RSS recognizes the usefulness of its reagents and analysis regime in genealogy or investigations of ethnic origin – not exactly big markets with legions of eager licensees, so why bother staking an IP claim? On the other hand, RSS may find that its reagents can be used in other commercially relevant sequencing techniques. Patenting these alternative uses may open additional markets (or prevent their loss to competitors).
- *Keep the patent fire burning.* A nearly universal principle of patent procedure restricts each patent application to a single invention; claims violating this rule are "restricted out" of the application, and must be pursued in separate, parallel "divisional" applications. The United States takes this idea a step further, allowing applicants to essentially refile an application, again in parallel with the original, with whatever claims are desired – so long as they are supported by the original application's text. These "continuation" applications, as well as divisionals, retain the original priority date of their parent applications.[2] Moreover, it is possible to file continuations or divisionals

[2] A related animal, also native only to the United States, is the "continuation-in-part" (CIP) application, which contains text from the earlier application as well as new material. These applications are often

of other continuations or divisionals (although the term of any resulting patent cannot exceed 20 years from the filing date of the earliest application in the chain). This creates strategic possibilities. Keeping a continuation or divisional pending, in reserve, allows its owner to react to new threats or opportunities appearing in the marketplace. If a competitor attempts to design around a patent by exploiting some verbal subtlety in the claims, the patent owner can file a continuation application with claims that wrap more tightly around the new product's throat.[3] If some peripheral use of the core technology does not initially justify patenting efforts, it can still be described in the application and claimed later in a continuation. At the very least, its publication will prevent others from gaining pre-emptive patent protection.

- *More is better.* All else being equal, more patents are usually better than fewer. It should not work that way, since patents are collections of claims and each claim should stand on its own, regardless of where it is found. As a practical matter, however, a judge or jury may be predisposed against any claim wedged in among others it has just found invalid. The foul stench of invalidity, in other words, can engulf an otherwise inoffensive claim, so survival chances tend to improve as the number of patents increases. And, as noted earlier, the remedy for infringement does not depend on the number of claims infringed; one is the same as many.[4] When Kodak fatefully elected to dip its toes into the instant-photography waters, notwithstanding the dense bed of patent mines Polaroid had laid, it eventually found itself hoisted for nearly a billion dollars – despite invalidating numerous Polaroid patents and, in the end, being found to infringe only 20 claims. The primary cost difference between one patent covering multiple inventions and a single patent for each invention is largely administrative, and may be inevitable anyway due to restriction practice. From a strategic perspective, therefore, a team of lean, focused patents is almost always preferable to a bloated loner.
- *Identify third-party rights.* Manufacturers as much as licensors must keep aware of others' rights, although for a different reason. Here it is a matter of discovering and mitigating potential problems before they assert themselves to devastating effect. A supplier rolling out a new product will be chagrined

used to cover improvements, receiving a split priority date: the filing date of the original application for material drawn from that application and the CIP's own filing date for new material.

[3] Thanks to some recent court decisions, a bit of uncertainty now pervades this strategy – but only a bit. It is likely to remain viable as long as United States law recognizes continuation practice.

[4] On a product-by-product basis, that is. For example, if fewer products infringe the narrower claims than infringe the broad claims, then invalidating the broad claims will reduce available damages. The point is that, if a given product infringes at all, it does not matter how many claims it infringes.

to find itself on the receiving end of a restraining order the day it hoped to ship. Its customers will be even more chagrined, and may never return. Inconvenient third-party IP positions can often be managed – for example, by designing around patents, by invaliditating them, or by licensing them. But only if they are discovered early.

The international dimension

International patenting efforts, as we have seen, proceed slowly and expensively. Still, because patent protection stops at the borders of each issuing country, any serious international sales or licensing efforts must consider the need for correlative patent coverage.

The goals of any foreign patent program must be, first, to identify countries in which the economic value of patent protection exceeds the cost of obtaining it, and then to prioritize among such countries within the available budget for foreign IP protection. The basic selection criteria should include:

1 *Cover prospective markets, not potential infringement sites.*
2 *Ensure that the subject matter is protectible in each country of interest.*
3 *Favor countries in which measurably substantial business is likely to materialize within 18 months, or in which realistic licensing partners reside.*
4 *Favor countries with satisfactory enforcement environments, and in which patent infringement can be detected.*

Let us take those criteria one by one. The idea of obtaining IP coverage in market countries, rather than where infringement is likely to take place, strikes some people as counterintuitive – like using fingers to plug a leaky dyke rather than doing something about the flood. After all, the source of the infringement is never attacked, and market coverage will inevitably be incomplete. The problem is that countries popular with infringers tend not to offer effective legal recourse against infringement; that is why they are popular. Taiwanese manufacturers, for example, are legendary for the speed and stealth with which they can set up assembly lines and, at the first hint of official sanction, relocate. India's judicial backlog and the epochal delays faced by would-be litigants are equally renowned. Moreover, infringers can bounce from country to country in this age of instant global communications. For all these reasons, it is typically far easier to block the borders of your primary markets than to hunt the elusive snark.

Excluding infringing subject matter from importation means enlisting the aid of customs authorities, and the procedures for doing so vary from country to country. In the United States, a US company may initiate proceedings against foreign infringers of patents, copyrights, or trademarks through the US International Trade Commission (ITC). The procedure is every bit as onerous as federal litigation, and in some ways more so. An ITC proceeding is part administrative investigation, part trial. First, the accuser must convince the ITC to initiate the case, specifying the precise nature of the infringement and demonstrating its harmful effect on a domestic industry. Once involved, the ITC – more specifically, its investigative staff – itself becomes a party to the action, and the case itself moves ahead relentlessly with the ITC in charge. The accuser has no brakes to step on; he cannot, for example, decide to abandon the action or settle. The proceeding includes the full range of discovery measures, which take place, along with hearings and trial, on a compressed schedule. Most investigations are fully resolved within a year.

Money damages are not available in an ITC proceeding – for those, a parallel case can be pursued in court – but the available remedies strike fear into infringers' hearts: exclusion orders, enforced by US Customs, bar entry of the goods into the United States. Moreover, in some cases the exclusion orders are "general" in nature, covering any source of the goods from a particular country rather than the specific infringers identified in the investigation.

For subject matter involving only registered copyrights and/or trademarks, but not patents, the US Customs Service offers a far less demanding avenue for blocking imports. The "recordal" procedure allows the IP owner to record its registrations with Customs and actively assist its agents in identifying, and excluding, infringing goods. This procedure is especially effective against high-profile counterfeit merchandise.

In other countries, customs authorities act against importation of infringing goods in response to a seizure request by the IP owner or following successful action in court. Matters grow complicated in Europe due to the continent's simultaneous (and occasionally contradictory) identities as a single market and a region of sovereign states. The Treaty of Rome, which underlies and largely defines the European Union, mandates the free flow of goods among European countries. What happens, however, if a product is patented in some European states but not others? For infringing products made outside Europe, the answer is easy: they can be blocked from entry into European states where patents would be infringed. For infringing products made inside Europe (that is, in a country where no patent has been filed) the question is more complicated – particularly if the patent owner has himself put the goods on the market in non-patent countries. This can happen, for example, if a company sells throughout Europe but has secured patents in only some European countries. It may be very difficult in such circumstances to reconcile pan-European marketing efforts with selective patent enforcement.

Keep in mind, too, that the criteria set forth above are just that: criteria, not fixed rules. The market-blocking strategy presumes a primary market consisting of a few countries and a wide open world where infringers may set up shop. That picture may be accurate, for example, in the case of RSS. Would-be competitors can set up a reagent laboratory virtually anywhere, but the juice will be useful only in countries wealthy enough to afford sophistocated sequencing equipment – the United States, Europe, and Japan will almost certainly constitute the vast bulk of the marketplace.

Does the reverse ever occur? Are there ever fewer territories of potential manufacture than purchase? That may well be the case for our friends at True Blue, from the first case study. If True Blue's wafers require high-end semiconductor fabrication facilities, which cost billions of dollars to construct, these simply will not exist outside a handful of countries (primarily the United States, Japan, China, and Singapore). The market for consumer electronic products that employ blue lasers, on the other hand, is fully global. If True Blue decides to pursue patent protection for its process recipe, it may well wish to keep manufacturing countries at the top of its foreign IP priority list, with consumer countries secondary. Unusual market structures call for altered strategies.

Pharmaceutical companies face the worst of all possible worlds – a wide-ranging marketplace and many countries capable of manufacturing drugs. No wonder they spend so much on worldwide patent protection.

Now for the second criterion. As noted in chapter 1, the patent laws of different countries reflect local policies. The United States has rolled out the welcome mat for business-method patents. Virtually all countries impose some restrictions on patenting (or enforcement of patents) covering medical techniques. In some countries, that ban extends all the way to pharmaceutical compositions – not just the use regimens. Divergence can also be seen in biotechnology areas, particularly those involving living organisms. The most liberal policy, once again, is found in the United States, where life forms shaped by human effort may be patented. And computer software, while theoretically excluded from the patent systems of most countries (except those liberals in the United States) if claimed directly, can in most cases be protected through artful drafting. But it is always important to check first, and often; patent laws evolve. Doing the research and planning an international strategy early, rather than simply reacting to the realities when it is too late to change business direction, will sidestep unexpected IP roadblocks and strengthen the overall prospects for success.

A virtually universal rule of patent eligibility is "utility," sometimes called "industrial applicability." The utility requirement bars perpetual motion machines because they cannot work (and are, therefore, useless by definition), but also research curiosities that lack direct commercial or industrial application. Usefulness in research is not enough; a patent is not a license to fish but a reward for the catch. The utility requirement looms especially large in the pharmaceutical area, where all potential new drugs are research curiosities until efficacy can be shown. Demonstrating a beneficial effect in animals may satisfy the utility requirement for veterinary applications, or even for human applications if sufficient correspondence between animal and human responsiveness has been established. Otherwise clinical data may be required.[5]

What about research tools, such as probes used to plumb the depths of DNA in order to locate genes or other areas of interest? Certainly a gene has utility. But what about the probe? Most patent offices view research tools as lacking the requisite utility unless they have independent diagnostic value – that is, can be used directly to detect genetic abnormalities or other conditions rather than merely as bait to fish in unknown DNA regions. So a probe employed to locate a full-length DNA sequence with a known and clinically significant function is probably patentable, but, if the DNA sequence is unknown or lacks such function, efforts to patent the probe will probably fail.

The third foreign-filing criterion – favoring countries where business is likely to materialize soon – can be considered a sanity check. Hope springs eternal in the human breast, but hope will not pay for those filing fees, translation costs, and annuities, which drain cash *today*. Be realistic about where, when, and to what extent sales are likely to occur. The PCT mechanism helpfully postpones the onset of significant costs, but filing too many PCT applications exhausts resources better directed toward development of domestic patent rights. Of course, the issue is one of emphasis rather than a stark choice. Foreign rights should be neither shunned dogmatically nor embraced unquestioningly. The pole star, as in all matters of IP strategy, is the direction planned for the business. A firm whose market strategy leans primarily toward licensing, for example, should attempt to view priorities through the eyes of its prospective licensees or acquirers, probably placing greater emphasis on foreign IP rights than a company struggling to establish itself as a manufacturer.

Similarly, while common sense dictates a preference for countries with favorable enforcement environments for IP, reality often mandates otherwise. Primary markets must be protected. The absence of effective recourse for IP violations in a particular country may influence the decision whether to do

[5] This can generally be submitted after filing, during the prosecution phase, without affecting the priority date; there is no reason to delay filing until clinical data can be obtained.

business there, but, once that decision has been taken, IP efforts must proceed – and the enforcement chips will fall where they may. On the other hand, when prioritizing IP efforts among different market countries, preference should go to those that support enforcement.

Let us now step back. We have discussed the factors favoring and disfavoring foreign protection as a matter of business strategy. Taking the opposite perspective, do any considerations affirmatively recommend a domestic-only policy? Many companies, particularly those whose primary market is the United States, content themselves with local rather than international protection for two reasons. First, the sheer scale of the US market for so many different products makes it almost impossible to forgo; just as US companies are well-advised to seek international markets, few foreign firms can achieve significant, sustained growth without sales in the United States. That means that an effective United States IP portfolio may dissuade foreign companies not only from entering the US market, but from competing altogether as a result. Second, as noted in chapter 1, US patent law sometimes has an extraterritorial reach, covering activity that occurs abroad but has an effect in the United States. Obviously that is not the same as having a foreign patent. But, if competitors are likely to access foreign markets from the United States, or if a foreign competitor's activities implicate the US in some way, these extraterritorial laws may be the next best thing.

Be forewarned: the analysis is a bit involved, and the faint-hearted reader may wish to skip ahead. But not if you plan to rely on these provisions. They are one ensemble you will want to try on carefully before buying; getting it wrong means going naked abroad.

The first extraterritorial provision, section 271(g) of the Patent Act,[6] makes it an act of infringement to import, sell, or use in the United States a product made outside the United States in accordance with a US-patented process. In other words, while a US process patent nominally covers only domestic use of the process, this provision prevents people from circumventing the patent by performing the process abroad but commercially exploiting its results in the United States.

Let us see how this would affect True Blue. They make wafers from which blue lasers may be fabricated, and the lasers themselves find their way into consumer and business products ranging from DVD players to mass storage devices. Let us say True Blue successfully patents its process recipes for creating crystals and then wafers. Let us also say that the United States market is so

[6] 35 USC section 271(g), to be precise.

large that walling it off from competitors would, as a practical matter, seriously impair their ability to get into the game.

What happens if a competitor in China makes wafers using True Blue's recipes and then turns them into blue lasers for sale all over the world? Seems like True Blue is in trouble, since all of the processing activity takes place abroad, beyond the reach of its United States patents. But, in fact, section 271(g) permits True Blue to march into the ITC and demand a halt to importation of the devices based on its recipe patents, effectively extending their effect to activity outside US borders. Section 271(g) does, however, contain a couple of exceptions. First, it will not cover products that are "materially changed by subsequent [unpatented] processes." Second, it will not cover a product that "becomes a trivial and nonessential component of another product."

Let us take that second exception first. If True Blue's competitor does not ship wafers into the United States, but instead markets them to Korean laser manufacturers who sell to US importers, this exception may operate against True Blue. It depends on the application of those two words "trivial" and "nonessential." As the heart of the laser, the wafer certainly is not trivial; but, if noninfringing substitutes are available, perhaps it's "nonessential."

Now suppose that True Blue's Chinese competitor sells lasers to a US company that uses them as what we will concede to be trivial and nonessential components of a complete DVD system. Section 271(g) can still be used to stop importation of the lasers, although not manufacture of the system; that is, True Blue can sue its Chinese competitor but, because of that second exception, not the US DVD manufacturer.

The other exception to section 271(g) means that, if True Blue's wafers are materially changed as they are fabricated into lasers, then importation of lasers will not constitute infringement. Of course, lots of things may happen to a wafer during device manufacture. But, if its basic structure persists in the finished device, there is a pretty good chance section 271(g) will still apply.

Courts have recently tacked on a third limitation, although it is not terribly relevant to True Blue. Section 271(g) applies to methods of manufacture but not to methods of use. It will not extend, for example, to methods of designing or discovering something – for example, if a US-patented method to screen for new drugs is employed abroad, section 271(g) cannot be used to stop US manufacture of the resulting pharmaceuticals. The US activity is too many steps removed from what is actually patented.

Let us now switch our facts around a bit. Suppose that, through the vagaries of patent prosecution, True Blue winds up with patents covering lasers that employ their wafers rather than protection for the wafers themselves (or for methods of making them). What happens if a US competitor were to export wafers, made using True Blue's *unpatented* techniques, for assembly abroad into lasers that its patents do cover – that is, *would* cover had those lasers been made or used in the United States. In this case, another extraterritorial patent provision – section 271(f) – may apply. This provision also covers situations in which otherwise infringing activity occurs outside rather than within the United States, but where products rather than processes are involved. Section 271(f) prevents competitors from selling, in the United States, unpatented components of a US-patented product so that they can be combined abroad into that product. Stated differently, this provision prevents avoidance of United States product patents through the domestic sale of noninfringing components destined for assembly outside the United States into the patented product.

The first part of section 271(f) covers inducement – that is, where the wafer seller brazenly encourages his customers to turn the wafers into lasers outside the United States. The second part covers instances where the seller is not quite that dumb. Those who sell even a single component of an invention, knowing and intending for it to be combined abroad into the patented product, fall within the reach of this second part of section 271(f). But, in order to shield legitimate activity, the provision contains some exceptions. First are the requirements of knowledge and intention. Of course, the more components of a product that the seller supplies, the more difficult it will be for him to feign innocence.

Second, the component(s) cannot be "a staple article or commodity of commerce suitable for substantial noninfringing use." Sellers of generic products like resistors and capacitors, for example, should not feel the sting of section 271(f) merely because someone uses them to build an infringing radio. But, if gallium arsenide wafers have no use other than to make True Blue's lasers, then exporting wafers will have the same legal effect as manufacturing the lasers in the United States.

In sum, if business goals are diffuse and foreign sales a long way off, one could do worse – at least in the United States – than an international strategy that begins and ends at home.

(Not so) petty patents

Many countries other than the United States maintain a secondary system of patents for inventions that do not meet the traditional patent criteria. Once called "petty" patents and now less judgmentally known as "utility models" or "innovation patents," these instruments provide some patent-like rights with far less fuss and delay. In some circumstances, such lesser grants can prove a worthy adjunct or alternative to conventional patents.

Utility models are offered in a number of European countries – no pan-European version is available, in contrast to standard patents – as well as in Japan, China, Korea, Taiwan, and other countries. They have certain advantages (particularly when the alternative is no patent protection whatsoever) that stem from a less demanding threshold of inventiveness. Whereas a standard patent application undergoes stringent examination against prior work and actually issues only if a relatively high invention bar is cleared, a utility model may require no more than novelty and usefulness. In some countries, such as Spain, only local publications can be cited against the application. As a result, utility models tend to be granted quickly – in weeks or months, as compared with years for a typical standard patent. (Indeed, a utility model may not even be examined other than for compliance with formal requirements.) Not surprisingly, they are relatively inexpensive.

On the other hand, this very ease of registration underlies the utility model's disadvantages. A lesser form of patent, the thinking goes, should have a shorter life and fewer teeth than the real thing. Most utility patents last 4 to 10 years; the term of a standard patent is typically closer to 20 years. Moreover, without the benefit of examination, utility models receive a chillier reception in court. Many countries also limit the eligible subject matter; a German utility model, for example, cannot cover methods or processes.

Australia's innovation patents may differ least, in terms of IP enforcement value, from traditional patents. An Australian innovation patent lasts for 8 years and is not restricted in terms of subject matter: anything that may be covered by a standard patent (with the exception of plants and animals) is likewise fair game under the innovation system. An invention qualifies for protection unless it differs from a single piece of prior literature or prior use only in ways that make no substantial contribution to the working of the invention. In other words, even a small but meaningful difference – meaningful, that is, in terms of the invention's operation – will suffice. By contrast, to qualify for a standard patent, an invention may be measured against the combined teaching of multiple prior references, and must differ in ways that would not be obvious to engineers or scientists in the relevant field. An innovation patent, therefore, can be used to protect modest innovations that would not qualify for a standard patent.

Examination on merits is necessary to make an innovation patent enforceable, but can be deferred indefinitely. Recent experience suggests that examination can be completed in a matter of months, compared with 2 to 4 years for standard patents. Yet, despite their rapid progress through examination and the lower inventiveness threshold, once examined and certified they should be no less effective in Australian courts than standard patents. A successful applicant receives rights akin to a standard patent together with the usual remedies – damages and, in principle, injunctions that put a halt to infringing activity. Moreover, because innovation patents do not require an inventive step, they will be harder to invalidate than standard patents.

Disappointed applicants for conventional patents often seek secondary patents as better-than-nothing solace. But their greatest strategic relevance can arise under different circumstances:

- *As a quick weapon against infringement.* Suppose a patent is pending when the applicant learns of a competitive product that would infringe the as-yet-unissued claims. By filing a utility model with narrow claims covering the product, but claiming the priority date of the pending patent application, the applicant can quickly obtain legally enforceable protection.
- *As insurance.* Standard patents and utility models often can coexist (provided the claims differ in scope). An applicant may wind up with a narrow standard patent as well as a broader utility model or vice versa. One may be less vulnerable against invalidity attack than the other.
- *Licensing negotiations.* When patents have yet to issue, a would-be licensor can find himself at a negotiating disadvantage; pending applications represent no more than requests for protection, and mere hope will be valued commensurately less than actual legal rights. By filing a utility model, particularly one literally covering products the licensee wishes to sell, the licensor can bring enforceable IP rights to the table. (Naturally it is important to provide, in the license, for payment of royalties on sales covered by utility models as well as patents.)

Establishing an identity

Thus far we have not said much about trademarks. Certainly establishing and maintaining a company identity must figure into any IP strategy. That said, innovation companies tend to rely far more on their technologies and product capabilities than marketplace recognition; the latter is expected to follow from the former. A Cisco customer may appreciate dealing with an eminent supplier, and Cisco's brand may well make the difference in a close purchasing decision, but ultimately what the customer wants is an affordable piece of equipment with certain specifications, thank you.

Still, although it may be folly to think a million-dollar Superbowl ad can substitute for market-responsive innovation, it is equally wrongheaded to ignore branding under the illusion that technology sells itself. Every business needs an identity. Most business owners understand this – in fact, may understand it too well. Few issues of company organization hold such emotional weight as the choice of a name. Any brand can acquire great value in the marketplace – occasionally, outside the world of technology, more value than any other form of IP. All of the trust and reputation a firm acquires ultimately gilds its name, which must be a worthy bearer of that wealth. Founders therefore rack their brains and get into fistfights over names that inspire, that are memorable, that project confidence and relevance to the business, that are easily spelled and pronounced. When at last they succeed, their IP lawyers will likely as not trash their cherished whimsies. To an IP lawyer, only three things are important:

- Is the name available?
- Is it protectable as a trademark? If so, is it strong?
- Can you get the domain name?

Name availability seems like an easy mechanical question, and in many countries it is – you just check the registry of corporate names and the trademark registry, and see if the name you want has already been claimed. Although the need to check in two separate places may seem odd, it is essential; corporate registration authorities and a country's trademark office perform different functions, and one office seldom knows what the other is doing. Many companies have happily incorporated under a clever name only to find they cannot do business under that name due to prior trademark rights; had they checked for trademark registrations first, they would have avoided this nasty surprise.[7]

Often, a literal search for the desired name proves insufficient. This is because of the nature of trademark rights. A trademark is a word, a logo, a number, a letter, a slogan, a sound, a color, or maybe even a smell that identifies a source of goods and/or services. The point about identifying the source is crucial. Trademark law does not protect creative name formulations for their own sake, but instead protects the public and businesses against marketplace

[7] The law distinguishes between a trade or business name, which identifies an enterprise, and a trademark, which indicates a source of goods or services. Trade names are, of course, usually used as trademarks; Xerox is both a trade name and an indication of source for duplicating equipment. For purposes of this discussion, we will not distinguish between trade names and trademarks.

confusion. The law does not care that someone pilfered your precious pearl of wit; it is concerned that the public will confuse the pilferer with you, to the public's detriment and yours. As a result, trademark law often covers more than use of the identical name. That means a newcomer may be obliged to create some distance between a new name and someone else's. Ultimately, it is the likelihood of confusion that governs how close is too close.

One important source of marketplace confusion derives from similarity of the businesses – the goods or services, the customers, the channels of trade. The more the businesses differ, the less different the name must be to avoid trademark problems. Often the identical name may be used for entirely different businesses. A second cause of confusion results from similarity of the marks. The law, seeking the semiotic factors that might lead a potential customer astray, considers similarities of *sound, appearance, connotation*, and *commercial impression*. Too much of any one, particularly in a similar line of business, can result in infringement of the earlier user's trademark rights – even if the name is spelled or pronounced differently. The wise business owner considers this question (particularly when it comes to sound-alikes) in all countries where business will likely arise; what sounds totally different in New York may be too close for comfort in Tokyo.[8]

Conducting a name search in the United States requires more work than in other countries. This is partly because US trademark rights generally derive from *use*, not solely from registration. In fact, a name that is not in actual commercial use cannot be registered at all. In most other countries, by contrast, registration is all that matters; he who wins the race to the trademark office wins ownership of the name. While it may be possible to have a long-unused mark stripped from the registry in such countries, its prior use by others is not relevant to initial registration.

In the United States, where a business can establish a trademark merely by using it, search tactics must extend beyond the records of the Trademark Office. A quick screening search using an Internet search engine, such as Google, should begin the process. Beyond that, many oysters can be opened in search of that pearl, including industry databases, corporate-name registries, "fictitious name" databases, and state trademark records (yes, trademarks can be registered at the state and federal levels). Searching companies exist to perform precisely these kinds of checks. The extent of the search effort depends on the degree of tolerable risk, and that typically follows from the importance

[8] Meaning is important in another respect. General Motors learned too late that its model name "Nova," spoken in Spanish, means "doesn't go."

of the name to the business. Heading the list, of course, is the name under which the company plans to do business.

The *strength* of a trademark often puzzles those steeped in marketing. IP lawyers love completely arbitrary, inherently meaningless names like Xerox or Kodak because they are strongest in the legal sense, receiving the widest legal protection. It is hard work to forge a brand identity based on fanciful names, of course, because the name itself does not educate consumers about the company's business and products. That effort falls to the sales and marketing forces, which tend to despise fanciful names for precisely this reason. But, once the hard work is done and consumers begin to associate the funky name with its originator, courts reward the trademark owner by according the mark broad scope – a competing name must be *very* unlike it to avoid infringement, and unauthorized uses even in unrelated lines of business can be actionable.[9]

How do courts make this assessment? What factors determine the proper level of protection? If we consider a completely arbitrary or whimsical name to fall at one end of a spectrum of distinctiveness, the opposite end is a completely generic word or phrase describing the business – Xerox at one end, "duplicating equipment" at the other. While arbitrary marks receive broad protection, generic words receive none: they cannot function as trademarks to distinguish a particular source of goods, because all sources produce what the words say; indeed, to allow any company to monopolize the vocabulary describing what it sells would be to put its competitors at a potentially fatal disadvantage.[10] The spectrum can be fleshed out a bit with intermediate categories:

Arbitrary ↔ Suggestive ↔ Descriptive ↔ *Merely* Descriptive ↔ Generic

"Suggestive" names evoke the businesses with which they are associated, but do not describe them; there's some mental leap involved. The name "True Blue," chosen to indicate a source of gallium nitride crystals and wafers for blue lasers, represents a suggestive name because there is nothing especially blue about the (transparent) crystals and wafers. Moreover, the play-on-words reference to the expression *true blue* – which *really* has nothing to do with semiconductors – further distinguishes the name, nudging it closer to the "arbitrary" side of suggestiveness.

[9] For example, even if confusion is unlikely, unauthorized use of a famous trademark may be prohibited as tending to "dilute" the mark's distinctiveness.

[10] But a trademark can itself lapse into genericness if it isn't policed effectively. That is why companies like Xerox go ballistic when people use its name as a verb or noun. Too much of that and the name passes into common parlance, losing its status as a trademark. The words aspirin, cellophane, and escalator, for example, once were trademarks.

Returning now to RSS, overlook for a moment, the fact that its full name, Ridiculously Small Samples, is a ridiculous moniker. On the trademark spectrum, it is certainly descriptive of the company's business, but not generic; after all, RSS does not sell samples. The name evokes the *capabilities* of the company's products, rather than the products themselves. Most descriptive names can be registered and receive trademark protection. But, depending on the jurisdiction, there may be a hurdle: in the United States, for example, a company must show persistent enough use of the name that the public has come to associate it with the company. The legal term for this is "secondary meaning" or "acquired distinctiveness." While arbitrary and suggestive marks are considered to be inherently distinctive, suggestive marks must work to attain trademark status.

What about the initials "RSS"? While letters, initials, abbreviations, and acronyms may sometimes qualify as arbitrary marks if they are uncommon and do not merely stand for a descriptive phrase, secondary meaning frequently will be necessary for trademark protection. If RSS hopes to build trademark rights around those letters, then, it will most likely need to do business as RSS (and not under its full corporate name) in order to foster the necessary association between its initials and its products in the minds of customers.

Companies must also consider the availability of an Internet domain name. The race to establish domain names has essentially emptied the entire English dictionary, so companies hoping for Internet addresses that roughly match their names often resort to cumbersome word concatenation (www.ridiculouslysmallsamples.com is a lot of typing) or completely artificial, often goofy names that make their marketing staffs cringe and their trademark lawyers rejoice. Owners of domain names have often clashed with prior users of those names as trademarks, and the opportunistic efforts of some to register others' trademarks as domain names for the sole purpose of extorting money have earned them the sobriquet "cybersquatters." Although the rarefied realm of Internet governance was slow to react, trademark owners can now seek redress through the Internet Corporation for Assigned Names and Numbers (ICANN) or in the US courts (since the domain registries are located in the United States).

Companies occasionally place their hopes in symbols or logos, but seldom can these serve as the basis for a business identity. Even well-known symbols such as the Nike swoosh and the Playboy bunny conjure their companies' names most immediately; in a sense they serve the name more than the business.

Trademarks, like patents, are typically registered on a country-by-country basis, although, also as in the case of patents, certain regional registration options are available. Virtually all countries provide for trademark registration, and in most places using the mark on an unregistered basis is a mistake. It is too easy for a competitor to register the name and take it away even from an earlier user. In the United States, where use reigns supreme, trademark registration is less essential but highly advisable. Registration gives the trademark nationwide effect, an avenue into federal court, and certain beneficial presumptions about the mark and its ownership. The cost of registration ranges from around $500 to $2,000 per mark and per country (whether pursued individually or in regions), depending on local fees and who prepares the application. The cost is well below that of a patent, and, given the ease with which names can be stolen, companies should be liberal with their foreign trademark efforts.

A typical trademark strategy, therefore, might go something like this:

- Select a strong company name.
- Search that name intensively, register it everywhere the company is likely to do business, and secure the domain name.
- Search every product name for trademark registrations and possible sound or connotation embarrassments in foreign countries, but skip intensive searches for all but the most prominent brands – and register at least those.

4 Implementation

Genius is 1% inspiration and 99% perspiration.

Thomas Alva Edison

Substitute *business success* for genius and this familiar aphorism applies with even greater force. When it comes to IP, the strategizing part is the 1% and the rest is implementation. But, after a seemingly sensible strategy has been put into place, corporate grandees may be tempted to pat themselves on the back and bow out, leaving the details to trusted vassals. It is not overstating matters to say this will probably doom the strategy – not only because 99% of the journey lies ahead, but because only through implementation can the strategy itself become refined and respond to ever-changing business circumstances.

Dialog is the current that drives this feedback loop. Central to effective decision-making in any enterprise, dialog means engagement rather than one group listening to its own voice. All key constituencies – upper management, marketing, and technical – must participate. In ways that make sense, of course: upper management need not get mired in the minutiae, nor should the hands-on IP administrators make all the decisions. How to tether the different players together and orchestrate their efforts into an integrated process is what implementation, and this chapter, are all about.

CASE STUDY #3: Few look forward to a trip to the dentist, and it's the special masochist who relishes a root canal. Mouthscape Corp. has a new system that can model a tooth and produce a crown, cap, or filling immediately, while the patient waits. It won't alter the basic procedure, but at least spares the patient the agony of a return trip. As the dentist scans a wand over the patient's teeth, Mouthscape's system builds up a 3D image of the entire dental anatomy. The dentist can inspect and manipulate the 3D image on a display screen, zeroing in on a problem tooth as the patient watches, soaring and swooping over the damage as he delivers the bad news. While the patient's credit card clears, the dentist tells the Mouthscape system to generate, say, a proposed crown. The system computes the configuration of the crown based not only on ideal tooth shape but also on the patient's actual bite, anticipating how the crown might grind against opposed teeth. Dentist and patient watch the display as the crown descends on to the image of the patient's unfortunate tooth. If unsatisfied with what he sees, the dentist may widen, shorten, or lengthen the crown as appropriate. Then he sends the complete model to a nearby carving machine that cuts the crown out of a small block of porcelain in a short time. Soon crown and patient are joined for life.

Wanda Dalrymple, Mouthscape's president, is terribly proud of her company's new $100,000 system. A dentist herself, she had been growing frustrated with sales of Mouthscape's more basic devices, which provide 3D dental images but no capacity for immediate prosthesis creation. Large competitors, including General Electric, also make systems that give dentists a 3D look into their patients' mouths. But none can actually use those images to automate prosthesis work.

Cognizant of the need for IP protection, Wanda put her husband Derek, the company's accountant, in charge of getting that job done. The diffident Derek Dalrymple wasn't prepared for the conflict he encountered. Mouthscape's small but growing marketing department struggled against their large competitors, who seemed to absorb Mouthscape's every new imaging feature into their own products as fast as Mouthscape could brag about it. The marketing group implored Derek to patent the company's proprietary image-processing technology and user features in an effort to establish some market exclusivity.

The engineers reacted with incredulity. "It's just Graphics 101," Derek was told of the imaging techniques, nothing you can patent, and as far as the user experience was concerned . . . well, how can you patent an experience? Derek definitely felt some *very negative energy* from the engineers. He put the matter on hold. He is studying it.

The prosthesis system is a different story; no one else has yet entered the market with a competitive product. The team who designed this system comes from outside the company, having been retained by Wanda as consultants. They present as a team, marching crisply through their PowerPoint slides, giving monthly progress demonstrations on their laptops. Wanda is beguiled and the equipment certainly works as they promised. But there always seems more to do – faster data handling, design of the "next-generation" carving machine. The consultants made the patent process easy and Mouthscape has already filed nine patent applications for the system they devised . . . or keep devising. Derek isn't sure. Everything happens so quickly – the consultants send in a flurry of patent disclosures, Wanda shakes her head in admiration, and soon there are more legal bills. The consultants seem to have *an excess of energy.*

Derek is in over his head and has asked us to help him impose some shape and direction on Mouthscape's IP efforts. What should we tell him?

Developing and maintaining an IP inventory

Our agenda for Derek must begin with invention recordkeeping procedures. A company without institutional measures for documenting inventions is blind to its own treasure; innovation, if invisible to those responsible for making IP decisions, might as well not be happening at all. Putting the necessary routines in place, however, and ensuring ongoing cooperation means facing down a phalanx of rolling eyeballs – those of the inventors who would rather not be bothered, the administrator who must chase the inventors and organize their disclosures, the committee that will evaluate them, and perhaps your own.

But it is important.

Though they will not help resolve questions of emphasis or company priorties, recordkeeping procedures provide the foundation on which any IP strategy must be built. Documenting invention is a familiar exercise in "first-to-invent" countries (pretty much only the United States and the Philippines) where priority goes to the first inventor rather than to the first to file a patent application. In fact, the differences between these systems is not all it is cracked up to be, and establishing ownership is important in *any* country. The reason is that an employment relationship does not guarantee ownership of inventions by the employer. That may seem counterintuitive. Certainly, one might assume, an employer that routinely obtains strong employment agreements in its favor should be safe; isn't it enough to have employees acknowledge inventions and related IP rights as company property?

Not necessarily. Employment agreements might not extend to ideas the employee pursues on her own, and in some jurisdictions even the strongest agreement isn't enough. A handful of states in the US, for example, have enacted "freedom-to-create" statutes that *preclude* employer ownership of inventions developed entirely on an employee's own time, using none of the employer's resources. Agreements to the contrary cannot be enforced. In Japan, even if a contract entitles the employer to ownership of an invention, the employee may be entitled to "reasonable remuneration" beyond his salary. And, if the Japanese employer offers incentive payments to employees when they invent, it may be too little; an employee is free to file suit for more money.

It is critical, therefore, for employers to obtain assignments of rights on an invention-by-invention basis – certainly when a patent application is filed, but ideally beforehand. In jurisdictions with inventor-compensation laws, earlier is always better for the employer, since an award given when an invention is nascent and its value speculative will appear, in a court's hindsight, far more reasonable than the same compensation offered after the invention has become a blockbuster.

On the copyright side, it is often a mistake to assume that software is a "work for hire" and therefore owned by whoever commissioned it – particularly if consultants are involved. In the United States, the definition of a work for hire is quite technical, and, if the relationship in question fails to meet the statutory criteria, copyright remains with the developer. The situation in such cases is analogous to buying artwork: the purchaser acquires the object but not the right to reproduce it. Without an agreement expressly assigning copyright, the right to reproduce or even modify the software may rest with the developer.[1] And remember that moral rights, if applicable, cannot be disclaimed or assigned. Depending on the country, these may irredeemably restrict the owner's ability to make changes to the software without the express consent of the original author.

Regardless of the IP protection mechanism involved, the ability to obtain timely assignments depends on awareness of innovation as it occurs. That is where documentation procedures come in.

Notebook procedures involving patentable inventions are most common in the United States, where priority contests can turn on who invented first. Almost painfully archaic, these procedures date from an era when all inventions originated in laboratories and all researchers used the same kind of notebook. Even worse, strict adherence to detail may, in the end, fail to influence the outcome of a priority contest; documenting an earlier "notebook date" is only the first of several tests an inventor must pass before winning priority. More on that below. For now, let us focus on the recordkeeping habits the law still expects.

- Notebooks should be permanently bound with numbered pages.
- After each experiment or documentation of an idea, and desirably following each day's entries, every page of the notebook should be *witnessed* and *dated* by at least two co-workers or others who understand the technology.
- Witnessing co-workers must read and fully comprehend the notebook entries, and ideally – and somewhat incongruously – are not part of the team working on the invention.
- Copies of other documents bearing on the progress of the invention, such as correspondence, development reports, receipts for materials used in research, test results, etc., should be retained.

The reason for all this cumbersome punctiliousness, of course, is that the temptation to fudge dates can be awfully strong with personal fortune and a company's future at stake. Memory is tricky and the law is suspicious,

[1] It is also worth noting that, while ownership of a work for hire remains permanently with the employer or commissioning party, in the United States copyright *transfers* may be reversed after 35 years by the creator – far too much time for this prerogative to be of routine concern in the world of software.

demanding objective evidence that is difficult to fabricate.[2] Still, paper notebooks and witness corroboration fit poorly into the routines and cultures of much technology these days – it is hard to imagine one of Mouthscape's software engineers recognizing the need to stop coding and start writing up what he did that day, much less enjoying the experience. Innovators may be understandably tempted to rely on e-mails or back-up data records in order to establish invention dates, but, because the law is slow in catching up to technological progress, such expedients may be open to attack as insufficiently trustworthy. Commercial software that simplifies and automates notebook practice is also available, but these programs still tend to adhere to the established procedures – precisely so the records they produce will pass legal muster.

Now, suppose an inventor maintains meticulous notebooks. Can she lose priority even if she has all her pages and corroborations in place? Yes – US patent law promises priority to those who not only invent first, but who also diligently pursue their inventions. Depending on the specific circumstances, and the law here is complex, the first inventor can lose priority reaching back to her notebook date if she puts the invention on a shelf – even for a little while. If she stops working on it, perhaps just for a few weeks, continuity is lost and the earliest available priority date may be when she resumes her work on the invention. Moreover, even if she is diligent through to completion of the invention but then waits too long to pursue patent protection, the chain is again broken; this time priority reverts to her application filing date – the same result as under a "first-to-file" regime. So it is important to document not just the "conception" of an invention, but also its progress up until the day a patent application is filed.

Notebook procedures document inventions; ***disclosure procedures*** make inventions visible to the company. Invention disclosures represent an inventory – a repository and record of the company's innovation. They provide the basis for patenting decisions, certainly, but also record the provenance of an invention and may substantiate later trade-secret claims. Departing employees, for example, will be hard-pressed to claim ownership of technology for which they have previously filled out a disclosure. So from the company's point of view, bring on the disclosures – the more the merrier.

From the perspective of the engineers and scientists who actually innovate, on the other hand, disclosures are tedious at best and torture at worst. They

[2] Inventors who work alone, outside a research community, have resorted to everything from notarized records to posting writeups of their ideas to themselves by registered mail and leaving the envelope sealed in order to establish a conception date. To obviate the need for these practices, the US Patent Office accepts disclosure documents for a nominal fee and retains them for two years.

subtract from otherwise productive research time. Energetic technologists constantly drive forward, while invention disclosures dwell in the past. Even innovators who recognize the value of IP would usually rather face the details *mañana*. So a company's disclosure program must balance the critical need for information with the research staff's capacity to provide it.

A more subtle issue is psychological. Engineers and scientists – unlike, say, their lawyers – are modest by nature, acutely aware that they and their accomplishments ride on the shoulders of the giants who preceded them. They may wrongly equate patentability with profundity. In fact, the standard is far more lenient – really more one of *un*patentability. Technologists must recognize patents as legal instruments rather than prizes for significant achievement. They should be encouraged to fill out disclosures liberally, without embarrassment over the seemingly mundane or trivial; disclosures must never be seen as pleas for professional recognition. So long as the company takes steps to minimize the paperwork burden (more on that below) and treats the process as a neutral recordkeeping function, it can expect its employees to supply a reasonably complete chronicle of the innovation occurring under its roof.

Establishing realistic procedures and responsibilities

A man who wishes to demand something difficult from another man must not conceive of the matter as a problem, but rather simply lay out his plan, as if it were the only possibility; when an objection or contradiction glimmers in the eye of his opponent, he must know how to break off the conversation quickly, leaving him no time.

Friedrich Nietzsche

A man without a plan is not a man – Nietzsche! (Big Boy Caprice, as played by Al Pacino in *Dick Tracy*)

Let us get back to Mouthscape and poor Derek. He must acquire some positive energy and take charge. As an organization, Mouthscape cries out for procedures that will empower Derek with information – a current and complete picture of Mouthscape's innovation – without overburdening the technical staff. But first everyone must be set rowing in the same direction, with a similar grasp of company objectives and appreciation for its IP priorities. The consultants understand IP but pilot their own maverick speedboat, with Derek floating behind in their wake. Some orderly, company-wide procedures, centering around Derek, will give him authority without seeming reproachful or diminishing the consultants' laudable (and very necessary) enthusiasm. The sullen engineers need to improve their IP IQ and accept some new responsibilities. Heeding Nietzsche, it is best to implement a new body of procedures all at once, so no group feels either privileged or targeted.

Derek should organize an IP seminar for all technical personnel, including the consultants, Wanda, and other Mouthscape executives. Part classroom and part revival meeting, such a seminar communicates a shared sense of mission in which IP plays a critical role. Sinners are forgiven their past transgressions and shown the righteous path – that is, the basics of IP protection outlined in chapter 1, but at a level of detail meaningful for the technical staff. A company's outside patent counsel often may be inveigled to make this presentation at no cost; their reward for such "missionary work," as it is often called, is a sounder sleep. The missionaries instruct the natives on patent tripwires, the importance of early disclosures, standards of patentability, the benefits and disadvantages of the various IP options, and common-sense practices involving proprietary information – for example, do not send anything remotely sensitive to anyone who has not signed the company's nondisclosure agreement.

Company leaders – that would be Wanda – deliver the homily, speaking to the pivotal role IP is expected to play in Mouthscape's success, stressing the value of everyone's ongoing participation. Wanda should explain the disclosure procedures, the need to put ego aside and err on the side of recording the banal, and, if the company has any patent incentives in place, now is the time to describe them. Such incentives may involve an award upon issuance of a patent or, if greater motivation is called for, escalating payments as progressive milestones (most typically, filing and patent allowance or issuance) are achieved.

IP seminars, like revival meetings, must be repeated periodically for their message to take root at an organizational level. Given typical turnover and, hopefully, growth in personnel, once a year should do it. When it is over, everyone can retire to their offices and cubicles refreshed, eager or at least resigned to implement what they have just learned.

As the administrator responsible for IP, Derek has several roles within this process:

- *Manage invention disclosures* by motivating inventors, screening what they submit, and making preliminary decisions about what to do next.
- *Serve as liaison to (and potentially serve on) the patent committee.*
- *Monitor performance* by continually reviewing the fit between the IP portfolio and the business strategy, and evaluating, to the extent possible, the market impact of the company's IP.

- *Serve as liaison to outside patent counsel.* That involves not only communication and bill review, but ongoing assessment of quality and assistance with literature searching and citation.

So who, exactly, is Derek – or more to the point, who is the ideal person to assume these various roles? He must understand the company's technology, in order to deal credibly with the research staff and understand their disclosures, but must also maintain a keen awareness of business goals. He must be knowledgeable enough to create invention procedures and possess adequate authority to enforce them. He must be respected by the patent committee, because, in a pinch, he may bypass them. He needs the executives' ear, since their participation is more important than they are likely to appreciate. He should understand basic accounting and be comfortable with numbers. And he needs a deft touch dealing with IP counsel, an outside law firm at first but later, perhaps, an in-house staff as well as needs grow – counsel whose enthusiastic efforts are critical but who must be replaced, augmented, or diversified as circumstances demand.

Quite an assortment of skills to ask of one individual – certainly of Derek, who today can claim but a fraction of them. He will have to grow in tandem with the company's needs. This is common; it is rare that a small company's initial team includes the ideal person, and, in any case, he will share responsibility with the patent committee. In more established firms, the position is usually filled at the vice-presidential level – the VP of research and development or product development, for example, with technical and marketing experience. He should report to the president and/or the chief technology officer. Derek already reports to the president in many ways, of course, and may well keep his job even as Mouthscape expands and outside investors insist on professional IP management. He has seen the company emerge from an embryonic state and has a feel for the competition and internal dynamics that no new hire can match. Let us hope the procedures we will help him put in place can stiffen his spine.

Invention disclosures. Technology companies too often view IP efforts as a process that starts with an invention disclosure and ends with decisions by the patent committee. That is a recipe for rigidity. Fixed procedures quickly grow sclerotic; it is flexibility that breeds cooperation.

Let us start with disclosures. They are important, as explained above, in documenting invention dates and recording the company's innovation assets. But researchers, even IP-savvy ones who respond to incentives, will instinctively postpone onerous obligations. If every researcher is expected to fill out a full-fledged disclosure form every time he or she does something innovative, it will be difficult indeed to persuade them to err on the side of disclosing.

Here is a typical disclosure form:

Invention disclosure

Please type relevant information within the space designated.
Potential inventors:
Descriptive title of invention:
Closest prior work *(attach publications if available)*:
Description of invention:
(a) Problem Solved by the Invention:
(b) Benefits of the Invention Compared to Prior Approaches:
(c) Potential Ways in Which Invention Differs from Prior Approaches:
(d) How the Invention Works (attach additional sheets if appropriate):

Other relevant information
When did you first think of this invention?
What record do you have to substantiate this date?
To whom did you first disclose this invention?
On what date did you make such a disclosure?
What written evidence do you have of this disclosure to others?
When did you first do any experimental work toward carrying out the invention?
Who observed the process of your experimental work?
When did you first make written description of this invention?

Publication and relevant art information
Please list any papers, abstracts, etc. describing the invention which have been published or submitted for publication. Include the title, journal, and date or estimated date of publication.

Please indicate whether any oral presentations (including slide or poster presentations) of the invention have been or will be made, the date and to whom.

Give details of any commercial use or offers to sell this invention.
NOTE: PRINTED PUBLICATIONS, ABSTRACT, ORAL PRESENTATIONS OR OFFERS TO SELL MAY RESULT IN IMMEDIATE LOSS

OF RIGHTS TO OBTAIN PATENT PROTECTION. PLEASE ATTACH A COPY OF ANY PAPER, ABSTRACT, OR OTHER PRINTED PUBLICATIONS, INCLUDING A ROUGH DRAFT IF PUBLICATION IS NOT YET IN FINAL FORM.

Commercial potential of this invention

Please summarize your assessment of the commercial potential of this invention (and note any products into which it might be incorporated):

INVENTOR(S) SIGNATURE		READ, UNDERSTOOD, AND WITNESSED	
____________	DATE	____________	DATE
____________	DATE	____________	DATE
____________	DATE	____________	DATE

This form has it all – everything Derek needs to understand an invention, how it differs from prior work, who invented it, the strength of notebook documentation, and whether patent-defeating events have occurred or are imminent. When completed, it will streamline the process of preparing a patent application. But it is absurd overkill if the objective is merely to build an IP inventory with minimal burden on inventors (not to mention Derek). If researchers document their work generously, as they should, reporting even modest developments, then only a fraction of what they describe will ever reach the patent office. They should fill out a much shorter disclosure form as a matter of routine – a form that will bring the basics to Derek's attention, alert him to any need for urgent action, and allow him to decide *whether to ask more questions or request the long form*, or instead merely to file the disclosure away for future reference. What seems routine or low-priority now may become exciting as it coalesces with other efforts or matures into a product. And Derek needs to know immediately about anything that is clearly compelling – while it is still a twinkle in its inventors' eyes – so he can have a quick provisional filed to establish the earliest possible priority date.

Inventors can be asked to answer the following undemanding questions, as a first step, whenever even a whiff of invention is in the air. They might even e-mail the information to Derek, who can print it out and, if he feels technically qualified to do so, sign and date the hard copy as a witness (and pass it over the desk for a second signature); Mouthscape's e-mail system, if routinely backed up and configured to certify transmission dates, further substantiates the date of conception.

Short-form invention disclosure

Potential inventors:
Problem solved by the invention:
Benefits of the invention compared to prior approaches, and how it potentially differs:
Brief description of how the invention works:
When did you first think of this invention?
Any commercial use or offers to sell this invention? If so, please explain.
Any upcoming disclosures of this invention? If so, when and where?

Still, there are those surly software engineers. If they shirk even this minimal obligation (and are valuable enough to Mouthscape to get away with it), the requirements can be reduced still further. The shortest invention disclosure of all asks but two questions – why and how? Any holdouts at this point need a new employer.

When a patent application – even a provisional – is filed, it is important to obtain a formal assignment of that application and any future ones drawing priority from it. The assignment should be "recorded," like a deed, with the patent office. If patents are not to be filed but software is involved, and there is even the slightest question whether any of the developers are formal employees of Mouthscape, Derek must obtain copyright assigments from everyone involved. Copyright assignments usually contain not only express ownership-transfer language but also characterize the subject matter as a "work-for-hire," whether or not that is really the case (on the theory that it can't hurt to try).

Dealing with the patent committee. Let us be optimistic and assume reasonably good inventor cooperation. What does Derek do with the steady rush of missives reaching his in-box? If Derek is a bureaucrat, he dutifully retains them until the next meeting of the patent committee. A smart IP administrator, however, reads every disclosure almost as soon as he gets it, makes sure he understands both the invention and its significance to his company, and fits it into his mental picture of the company's technology portfolio. He understands the company's business objectives and can set priorities. Ideally, the patent committee charts the flight plan, but the patent administrator flies the plane. What that means in practice depends on how the committee views its mission and how much responsibility Derek can handle.

The patent committee's job is to oversee IP development so the company's portfolio falls in the sweet spot of figure 3.1, and to ensure that resources are well directed. Patents and pending applications, particularly outside the United States, cost money every day they remain alive. Merciful euthanasia is called for when the technology they cover gets superseded, or if, despite all efforts, the coverage offered by the patent office falls below what is minimally acceptable. The detail involved in these decisions may tempt the committee to set a broad policy, fix a budget, and leave all the navigating to Derek. That shortchanges the purpose of the committee. The company only benefits when committee members take an active role and involve themselves, to some extent, with every application.

Who makes up this esteemed body known as the patent committee? In a small company like Mouthscape, it could very well be staffed by Derek, Wanda, the head of marketing, and perhaps Mouthscape's outside patent lawyer. It may seem pointless to schedule formal meetings among a closeknit crew who chat by the water cooler, but it is a good practice, one that professional outside investors often insist on. Regular meetings on a quarterly basis allow the company to review its strategy and approve major expenditures with all viewpoints represented.

The documents that set each meeting's agenda are the status list of the company's patents and applications (typically prepared by outside counsel) and recent invention disclosures received by the patent administrator. Patent status lists prepared by attorneys typically take the form of a table, with each row corresponding to a particular application and the columns defining status information: docket number, country, title, inventors, filing date and serial number, status, next action due. Sometimes issued patents occupy a separate list, since the only remaining strategic issue is whether to keep them in force. To the basic status list, the patent administrator must add two columns that only company insiders can complete: *business category* and *strategic objective.* Unless a patent's purpose is tactical, for example, setting up a roadblock for competitors or establishing an early priority date for new technology, it should be associated with a company product or at least a line of business – whether on sale or under development. The strategy column adds the necessary clarification, briefly detailing the business rationale underlying each case, and must be reconsidered frequently. A patent whose purpose is to cover a phased-out product is ripe for the dustheap, but if new strategic value has been identified – for example, coverage of a competitor's product or the potential for licensing – it may deserve to live another day. The status list must make all this apparent (see Table 4.1).

Table 4.1 Status table for RSS patents and applications

Attorney Docket	Country	Filed/ Serial No.	Inventor(s)	Title	Status	Business Category	Strategic Objective
MTH-001	USA	1/15/05	Paul Bustamante Rita Dorfman	Method and Apparatus for Rendering and Automatically Fabricating Dental Prostheses	Awaiting examination	Prosthesis design; prosthesis fabrication	Exclude competitors from pioneer product
MTH-002	USA	1/15/05	Jeffrey Watt	Carving Apparatus	Awaiting examination	Prosthesis fabrication	Cover best commercial practice

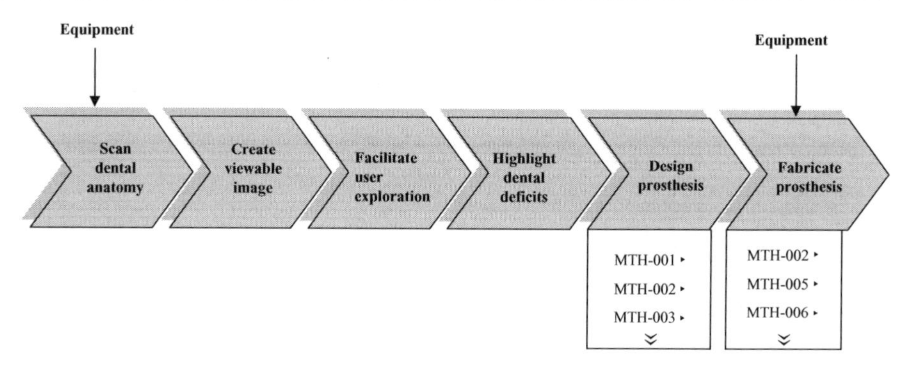

Figure 4.1

Derek might also consider preparing an interactive graphic relating Mouthscape's patents to its value chain (see figure 4.1).

Clicking on any of the value-chain elements produces a drop-down list of associated patent applications, as well as important disclosures that will soon become patent applications. This approach visually connects IP to specific business functions, and provides an immediate sense of which business segments have been most successful, or received the greatest attention, in terms of IP procurement.

Finally, Derek might consider preparing a table that relates patent coverage to competitor products and industry trends. This takes a substantial degree of technical and market sophistication – more than Derek has at the moment. Going through the exercise, however, with support from the necessary personnel can pay handsome dividends. This is how bottlenecks are discovered – gatekeeping technologies or business processes that competitors need to enter the market. Recognize such opportunities early, sew them up with patents, and those competitors will feel the pain. Maybe there is just one good – that is, fast or computationally tractable – way to scan and model the complex but predictable features of a tooth; maybe medical imaging systems are moving toward XML data representations; maybe Mouthscape's lawyer recommends logging a dentist's modifications to system-generated prosthesis designs in order to reduce liability exposure. Such areas may not stimulate much excitement from a *technology* perspective, but as a business matter an IP toehold could prove invaluable.

This exercise is particularly important for pure research companies that sell market positioning through licenses. Such companies should always attempt to stay at least one step ahead of their licensees, who are also, potentially, competitors. Few developers can sit back and milk a single early innovation like an annuity. Equally rare, however, is the innovator who can keep up with every competitor or maintain a leadership position across all applications of a technology. It is essential to pick and choose. Deciding where best to focus development efforts requires awareness not only of emerging trends and improvements, but also of their originators. The efforts and needs of market leaders cannot be ignored; their priorities may dictate research directions, even if they do not seem terribly inspired or cutting-edge. Commercial research, unlike its academic counterpart, is harnessed to the market.

Based on all of this information, Mouthscape's patent committee decides whether to turn provisionals into nonprovisionals, whether and where to file abroad, and when it is time to give up on an application or patent. The committee considers licensing offers and standard-setting opportunities, industry trends and the competition's latest products, infringement threats and brewing lawsuits. As for invention disclosures, Derek acts as a filter, essentially setting the agenda on this subject. The committee considers whether to advance or shelve various disclosures, whether to file provisionals or move directly to non-provisionals (hopefully after performing patentability searches), and, more generally, the company's IP direction and whether the patent mix remains sensible.

Lest all this sound deadly dull, patent-committee meetings can and should be fairly lively, even spirited affairs – if they are done right, if the proper people participate. Imagine the reaction when Derek shows figure 4.1 to the committee. Mouthscape's head of marketing may decry the exclusive emphasis on the prosthesis system; he is eager to see more patents across the board, covering every product he is expected to sell. Derek responds with the software crew's doubts. Now Mouthscape's patent counsel pipes up, voicing doubts on the doubters. Having often seen patent-averse engineers attempt to evade the process by declaring everything known or obvious, she tells Derek that curt dismissals won't do. Patentability is a judgment we should make together, she says. Let the software engineers explain what they've accomplished and show us specific examples of earlier work that anticipates their efforts; I'll be surprised if they have any. And, if you want, she continues, we can perform our own patentability search to dispel doubts further. Derek is intrigued, finally perceiving a rejoinder to all that negative energy. Wanda reiterates her love for the prosthesis system and its healthy mouthful of patent applications. The head of marketing groans audibly (but respectfully).

Let them clash (in a civil manner, of course). The business and technical sides can't and shouldn't always see eye-to-eye. Look again at figure 3.1 – the business folks' natural habitat is the left-hand oval, while the technical team lives for the right-hand side. Marketing often sees business opportunities where no patents exist; why the deficiency? Techies disdain the pedestrian and relish what fascinates; are their priorities wrong or is the marketing staff ignoring valuable opportunities? Thrashing it all out will help delineate the IP/business sweet spot.

Not everything can await quarterly meetings. A hot invention, a cease-and-desist letter from a competitor alleging patent infringement, trade shows, and unexpected sales – all of these events demand immediate attention. When an impromptu meeting cannot be scheduled, e-mail can stitch disparate schedules together, albeit imperfectly, at the price of less effective communication. If Derek is trusted enough within the company, perhaps he will have the authority to make certain decisions on his own, depending on circumstance. Often, for example, the patent administrator can authorize preparation and filing of provisional patent applications – particularly if the inventors have demonstrated willingness to shoulder a good part of the burden to reduce costs, and especially in emergencies.

Monitoring performance. The information presented in figure 4.1 and table 4.1 provides a pretty good picture of Mouthscape's IP portfolio and how it relates to the business. But hold on, Derek – there's more to your job than keeping the IP ship afloat. You need to measure its performance as well.

Performance has two components. *Internal* performance relates to the quality and quantity of participation in the IP effort (at the division, group, and individual level) and how much value Mouthscape receives for its IP dollar. *External* performance measures Mouthscape's IP program against its effect on the company's market position.

Monitoring internal performance requires some additional database tricks. When Derek receives a disclosure, he should log it into a "performance" database with fields for the date, the technology a disclosure covers, the identities of the inventors and their research groups, the disclosure's current status (on hold, superseded, patent application filed, etc.), and whether a corresponding product is being marketed. Figure 4.1 suggests how Mouthscape's IP relates to its business functions, but provides only an imperfect window into who is contributing and who is not. The entries in Derek's performance database can be sorted to see which groups are producing the most disclosures, and if any individuals within those groups seems especially prolific. This may reveal wellsprings of creativity or cesspools of egomania, depending on the quality of the disclosures and whether they are turning into patents and company products. Incentives, whether financial or psychic, can work too well; some researchers may seek to dignify or justify or magnify their efforts by filling out disclosures. They shouldn't be criticized too quickly; after all, most companies suffer from too few disclosures rather than too many. But the observation is useful to group directors and, ultimately, to the chief technology officer, who must be ready to trim or eliminate unproductive programs – for example, those that deliver many more disclosures than marketable products.

Assessing external performance is easy for a licensing company – which patents have been licensed and how much in royalties do they generate? – and almost impossible for product companies that do not license. Who knows, for example, whether a niche player's longtime market exclusivity stems from the strength of its patents or something structural about the market – barriers to entry, pessimism over growth potential, profits that satisfy one player but fail to attract competitors – or its good name. Still, there are indicators out there, direct and indirect, capable of providing insight. Are other industry players accumulating relevant patents? Their numbers and rate of increase suggest the intensity of competitive research. Are other patents citing your patents and published applications? In the United States, every patent lists the literature considered by the examiner, and these lists are searchable on the US PTO web site; it is easy to determine whether your patents are sufficiently relevant to others to merit citation. When they are, you might discover a competitor or even an infringer – or a patent you may be infringing!

Most companies do not have an entire market to themselves, however, and can develop at least a gestalt sense of which patents are working for them. They are the ones covering features that confer a marketplace advantage. They are the ones competitors ask to license or threaten to have re-examined.[3] If a patent covers a feature that makes a sale, or at least covers a product that customers buy for whatever reason, then a value can be ascribed to the patent. Not an especially precise value – as we will see in chapter 7, patent valuation can be

[3] If they are smart, they only threaten – we will see why in Chapter 5.

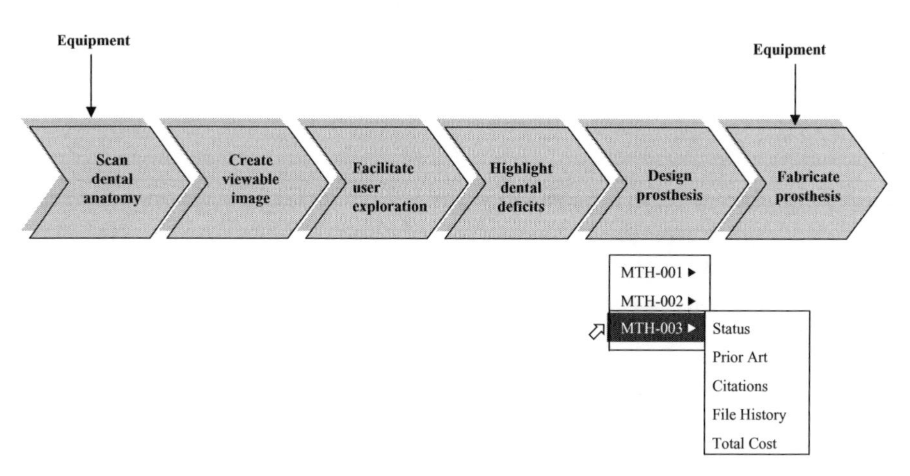

Figure 4.2

a tricky business – but a value that distinguishes it from its nonperforming brethren. That value can be compared with the patent's cost: total expenditures to date, plus the net present value of expected future costs.

So Derek may want to enhance the graphic shown in figure 4.2 to add some fancier database capabilities.

Highlighting an entry produces a menu of options. Selecting "Citations" provides a list of patents citing the MTH-003 patent or published application; selecting "Total Cost" calls up a spreadsheet detailing all costs to date, as well as projected future costs, for the case. Derek's polymath administrative assistant has set up the necessary database and updates the citation and cost fields on a monthly basis, dutifully searching the patent office's citation records and begging the accounting department (that is, Derek, wearing a different hat) to update the spreadsheet.

Selecting "Status" displays the row from table 4.1 corresponding to MTH-003. Assessing the strength of a patent may require review of its prosecution history – were severe compromises made with the patent examiner to secure allowance? – so Derek may want convenient access to this all-important paperwork. At least for recently filed applications, the US and European patent offices make file histories available on their web sites. Selecting "File History" runs a script that accesses and interacts with the US PTO web site to obtain the page showing the file history of MTH-003, or simply retrieves a stored image of the papers.

Disclosures are good, patents are better, and profitable patents are best of all. The ability to track all of these and, more importantly, to discriminate among them makes Derek the indispensable man, the go-to guy for Mouthscape's portfolio development. If he applies sound judgment to all the information he has at his fingertips, he will find the patent committee in frequent agreement with his recommendations and even reluctant participants yielding to the requirements he sets.

Dealing with outside patent counsel. Few companies and far fewer start-ups build their patent portfolios without the help of outside patent lawyers. Working with them effectively involves several distinct layers of management – all of which, while ultimately related, call for different judgments and skills. They include knowing when to call, establishing a budget and controlling costs, managing literature citations, and, most importantly, evaluating the substantive work product.

At what point in an invention's life cycle should the patent process begin? Surprisingly to many, it depends far more on the invention's importance than its stage of development. Even experienced patent administrators often fail to appreciate how little information a patent lawyer needs to prepare a patent application, and, as a result, applications often get filed – and priority dates established – far later than they should. While it is true that the patent system usually protects an idea's realization rather than the idea itself, expressing concepts in a practical format is part of the patent lawyer's role. If she has the proper training in the company's technology and the cooperation of the inventors, a patent lawyer can usually fashion a provisional or even a full patent application based on a telephone conversation or back-of-the-envelope description.

Not that it is always appropriate to do so. The less revolutionary the idea, the more prudent it is to await proof of concept – to see whether it can be realistically implemented as a commercial product. But, in the meantime, competitors are plying their own research efforts and filing for patents. Tick, tick, tick . . . excessive caution risks losing priority, while a quick trigger finger is guaranteed to bust the budget. To make this decision on a sound basis, a clear sense of the development timing is essential. Ask the inventors how long, really, it will take to (a) create a prototype, (b) begin beta testing, (c) establish a budget for "productization" and scale-up, and (d) achieve market introduction. Any of these events may serve as an informal deadline for action. A working prototype demonstrates the idea's viability, usually justifying the expense of a patent filing. But, if beta partners perform testing on a confidential basis and do not pay for the prototype (so that no eligibility-defeating "sale" occurs), patent rights may be preserved throughout the entire beta period. Budgeting for product scale-up means the company is firmly behind the project; there is certainly no reason to delay at this point, but so long as the entire development effort has remained secret – no public announcements, nothing on the company's web site, no nonconfidential demonstrations to customers or hints dropped at trade shows – a patent filing can theoretically wait until the day before shipment.

That would be crazy, of course, given the availability of low-cost provisional patent applications. Just how low-cost depends on the working relationship you establish with your patent attorney, and that relationship must be such that you do not tremble at the cost of picking up the phone and discussing strategy or asking for a provisional. Heavy users of patent services, particularly universities and large technology companies, attempt to manage costs by establishing ground rules concerning administrative charges and overhead – we do not pay for copying or postage, we do not pay for secretarial overtime, we do not pay for reporting or reminder letters, here are the maximum hourly attorney rates we will pay.

That bureaucratic approach misses the great green forest in the rush to cut down a few conspicuous trees. The most important aspect of cost control is control, not raw cost. Your legal bill equals your lawyer's billing rate times the number of hours she spends on your work, and the latter always has a far greater impact than the former. The more time she spends, the more you pay; the more your technical staff can contribute to the effort, the less you *should* pay, although it does not always work that way. Many attorneys simply refuse, consciously or unconsciously, to allow their clients' contributions to alter how they go about their work.

Those are the wrong lawyers to hire. Part of your patent attorney's job is to teach your technical staff, in the course of working with them, how to contribute effectively – that is, in a manner that replaces rather than supplements attorney time. This does not mean your engineers will write patent applications; they have neither the time nor the skills. But it does mean they can learn to write the heart of a provisional, and, when the time comes for the full-blown utility application, provide a rich enough detailed description so the patent lawyer can get the words mostly right the first time, without expensive redrafting and iterative exchanges. The longer a good lawyer and technical staff work together, the more of a rhythm they will establish, and the more costs will decrease.

Not that you will want a lawyer who needs much handholding in order to do his job. There will be times when researchers' time simply cannot be spared. Your attorney must be able to to write and file patent applications on the barest of disclosures, and at a moment's notice; those are usually the most expensive efforts, but the cost of tying up technical personnel on the eve of a trade show or product ship can be far greater. The point is that you, and not your attorney, must own the trade-off between time and cost.

Other points of practice style also affect the bottom line far more significantly than nominal rates and charges. An important one relates to handling rejections issued by the patent office. Your attorney is, above all, an advocate – not an addition to the research staff, more than a high-priced scrivener processing technical ideas into readable prose. Responding to patent-office rejections can cost $500 to $5,000, depending on their complexity. That is an excellent reason in itself to minimize the number of written responses; another reason is that every paper, every extra word in the file history can ultimately limit the resulting patent's scope. Patent lawyers should, therefore, endeavor to conduct a telephone interview with the patent examiner before filing a response. Sometimes it is hard to learn what is really bothering an examiner from the words of her rejection; sometimes examiners do not pay as much attention to a written argument as they should. Human-to-human communication forces both sides to flesh out and face all the issues. Simply "mailing it in," without that preliminary effort to achieve a meeting of the minds, is guaranteed to produce a lower success rate – translating into more written responses, increased time before patent issuance, and far greater cost.

In sum, cost control begins by getting it right the first time: during drafting, during prosecution, and also before getting started. A lawyer's awareness of strategic criteria and eligibility restrictions is just as critical as his practice style. Your attorney should help you decide between patent and trade-secret protection. He should also make sure you do not file for coverage of ineligible subject matter, which, once again, varies from country to country and over time. (While that sounds rather elementary, it is true that very expensive and inescapably doomed foreign applications are filed every day because no one performed the necessary review.)

Your attorney's performance in these areas can be measured quantitatively – cost per patent, average pendency period, etc. Obviously many factors outside your lawyer's control affect these parameters, such as the backlog at the patent office (which varies dramatically across technology segments), the complexity of the technology, and inventor cooperation. But over time, with enough data, a statistical picture will emerge. Far less amenable to objective assessment is the quality of the work. This can be extremely difficult to evaluate; patents are legal instruments, you might think, and I am spending good money to make sure my lawyers do them right! Well, your lawyers may be doing their best, but ultimately it is the company's job to decide whether their best is good enough.

There is a temptation to avoid asking that question. To their owners, patents, like children, are all so wonderful – strong, broad, articulate, everything a proud progenitor could ask for. The reality is often otherwise, unfortunately, because the process can easily go awry. A client with a new idea wants it protected in a hurry – often as cheaply as possible. His lawyer or patent agent has a certain level of skill, experience, and familiarity with the technology at issue. To keep costs low, he may not do a "deep dive" into the subject matter, particularly if it is unfamiliar (or so familiar it seems trivial), instead relying on the inventor's description. The inventor, for his part, may not have the time, verbal facility, or sense of what the lawyer is really after to write a complete exposition. The result, in such circumstances, is likely to be an unfocused document with claims that are too narrow or, worse, misdirected. It may be cheap or it may be expensive; cost is not a reliable indicator of quality. And someone like Derek, who has never litigated a patent, is likely to assume it is fine: it's a lawyer's document, once again, and he is paying plenty.

So how does Derek evaluate the quality of legal efforts? How can he tell a good patent from a bad one?

First of all, he should be able to understand it. As a matter of law, a patent is written not for Derek, but instead for "one of ordinary skill in the art" – an engineer or scientist. It need not be understandable to those of lesser skill, and, knowing this, Derek may be tempted to shrug and figure he does not comprehend his patents because he is not supposed to; he is not within the circle of knowledge. But a good patent reaches beyond the demands of the law. If Derek, steeped in Mouthscape's products, if not their theory of operation, can not follow his own company's patents, then how will a judge or jury? Certainly Derek will encounter technical arcana and unfamiliar terminology as the discussion descends into the details, but even here a good patent's coherence and linear explanation should keep it within his grasp.

Part of the art of advocacy is knowing how to tell a tale, and advocacy is ultimately what Derek pays his lawyers to deliver. Every patent must evoke a problem long endured until triumphantly solved by the invention. And that story must, at some level, align with the company's business plan. After reviewing the background and summary sections of a patent, Derek should have a strong appreciation for why the invention is important and how it works. That does not mean he receives a complete education on the technology and its underlying theory, but enough explanation to follow what the inventors have accomplished. (Too much education means expensive tuition – both in terms of legal charges for drafting the application and high translation costs when it is filed in foreign countries.)

The need for a strong story furnishes yet another reason to prefer more to fewer patents: the narrower the focus of the patent, the sharper its story will be. A diffuse patent covering numerous inventions can not do real justice to any of them.

The hardest part of a patent for anyone to read is the claims. Tough as the slogging may get, Derek and the inventors named on an application must develop an understanding of what the claims mean and what they cover; they are *their* claims, after all. Writing them is not a lawyer's exercise so much as a shared enterprise. The inventors must satisfy *themselves* that the claims cover the basic idea, foreseeable commercial products, and possible workarounds by competitors. If issued, the claims should leave those competitors frustrated and envious. None of this can be gleaned from a superficial reading, perhaps not even after repeated headbanging against those refractory words and clauses. Derek should talk to his lawyer and have her walk him through her reasoning and listen to his concerns and priorities.

Most patents should contain claims both to methods and products (or machines), even if the invention seems to be "about" one rather than the other. The reason is strategic, both in terms of coverage and enforcement. Almost any invention can be cast as a method or a thing; that is just a matter of patent drafting. Let us take the second application listed in table 4.1, entitled "Carving Apparatus." Sure sounds like a thing. Why bother to characterize Mouthscape's apparatus for creating pretty teeth out of hunks of porcelain as a method? The reason is that method claims need not recite a machine's component parts and can therefore reach closer to its theory of operation. That creates broader coverage; competitors' machines that do not have particular parts, but nonethless implement the guiding principles, will fall within the method claims – if they are drafted properly.[4]

Conversely, new processes – such as a fancy algorithm for generating dental prosthesis designs – should also be claimed in terms of generic apparatus and software for implementing them. Why? Think about who, exactly, will infringe the method claim. Probably not the competitor, but its customers. (The competitor just makes machines; it is the customer operating a machine who causes the fancy algorithm to execute – and therefore to infringe a method patent.) Certainly the competitor may bear *indirect* responsibility for

[4] That means method claims should not slavishly restate the elements of apparatus claims in active terms. They should recite operations rather than the behaviors of specific parts. Method claims should also be drafted to cover the activities of a single party. If it takes two to carry out all the method steps, for example, it is possible that no one infringes.

its customers' infringements, but a patent owner seeking recovery for indirect infringement of a method claim usually must demonstrate that the competitor overtly *induced* the customer to infringe. Sometimes a clever copycat can wriggle out of liability through, for example, ambiguous customer instructions, or at least raise an issue of fact that must be resolved at trial. But, if the software or hardware that the competitor sells can be covered directly (with an apparatus claim), there is no need to demonstrate inducement.

It goes without saying that a company must tailor its patents to the products it sells rather than the other way around. Yet, because patents can be so important to the outside world – to potential licensees, acquirors, and investors, not to mention aggressive competitors you would like to intimidate – it may be tempting to survey what patent coverage is available and herd development efforts into that open territory. That is almost always a futile exercise. What is important is best practice, that is, what the market will prefer and what competitors are likely to pursue if unchecked by IP rights; adopting second-rate technology merely so it can be patented invites improvement that skirts your patents and wins your customers. So, if your patents do not cover the best practice, change them so they do. If your patents *cannot* cover the best practice, it is time to rethink the strategic place of patents in your IP planning.

Dealing with outside patent counsel also involves managing literature citations. Most of this is technical and ministerial, but may save a patent from ignominious defeat in court. Particularly in the United States, every reference considered by the patent examiner before she agrees to allow the application gives the resulting patent extra strength; for reasons we will explore in chapter 5, it is very difficult for enemies of a patent to challenge its validity based on already-considered references. That is the carrot encouraging citations. The stick, wielded in the United States and an increasing number of other countries, is a penalty for withholding relevant information. Applicants for a US patent have a legal duty of candor toward the PTO. Breach that duty by failing to cite known, relevant prior literature and the patent may be invalid.[5]

As a patent portfolio grows, it can become difficult to ensure that all literature references cited in one case make their way into cases involving similar subject matter. Patent lawyers routinely cross-cite references among related applications within a "family," but not necessarily among applications that do not share a formal priority relationship. Unfortunately, even an innocent

[5] If the reference is directly on point, of course, the patent may be invalid anyway. But the draconian penalty of patent unenforceability due to "inequitable conduct" before the patent office may be applied even if the reference would not, in itself, have killed the patent.

oversight can be demonized long after the fact in the hindsight world of litigation. It would serve Derek well, therefore, to help in this process.

When a new patent application is filed, it can be compared with existing filings for similarity of subject matter, and the citation lists of all applications identified as similar combined for the new one. As the patent portfolio expands, of course, it may become onerous to assess each new application against every existing case. A partial solution is to ensure that all applications associated with a particular value-chain element share the same citation list (that is, when a new reference is cited against one application, it is immediately propagated to the others). The energetic database manager may take matters further by storing, in each patent application's database record, a paragraph summarizing its subject matter that can be searched by key word. If a new application involves carving, for example, Derek can use this word as a search term to locate related applications for their citation lists. Seem compulsive? You bet it does. But the more meticulous the citation practice, the less vulnerable the portfolio will be in litigation.

Managing participation in industry standard-setting

Suppose Mouthscape develops a form of digital image representation that allows its navigable 3D images to be displayed on conventional web browsers and media viewers; maybe you cannot soar and swoop, as Mouthscape's proprietary system allows, but at least static images of teeth and crowns can be viewed on digital X-ray systems already in widespread use by dentists.

The universal availability of this approach to image representation would be a good thing for Mouthscape, making its system compatible with existing equipment and thereby lowering customer resistance to adoption. To encourage developers of browsers and media players to support its imaging convention, Mouthscape wants not only to give it away, but to have it approved as an industry standard. It believes that the prestige and publicity associated with a standard will stimulate industry-wide acceptance.

Should Mouthscape merely publish its imaging convention, or patent it as well? Is IP even compatible with industry standards?

Let us take the second question first. At first blush, the idea of a "patented standard" seems oxymoronic. But today virtually all standards organizations have guidelines permitting members to contribute proprietary technology. Most state the same conditions: the contributor must disclose relevant patents or patent applications before its technology is considered, and must agree to grant licenses in accordance with policies established by the standards organization. While some require contributors to grant royalty-free licenses to anyone wishing to adopt the standard, most will tolerate fee-based licensing that is reasonable and nondiscriminatory. More on that in chapter 7.

In the United States, companies that fail to adhere to these guidelines may face the wrath of the Federal Trade Commission (FTC), which has treated violations as unfair competition. Recent subjects of that administrative wrath include Rambus Inc. and Dell Computer Corp. For example, the FTC found that Dell had participated in standard-setting activities concerning the VL bus (which carries information between a computer's central processing unit and its peripheral devices), and as part of the process certified it had no patents covering the standard. Later Dell tried to enforce just such a patent. The FTC issued an order that stripped Dell of its ability to do so.

Competitors may also file private lawsuits. Enforcing a patent withheld from a standards body against adopters of the standard can qualify as "patent misuse" or, depending upon the circumstances, constitute an antitrust violation.

Even if Mouthscape intends to contribute its IP on a no-royalty basis, it must still learn and follow the intellectual property guidelines published by whichever standards organization it works with. That means more work for Derek. As Mouthscape's patent portfolio grows, Derek must maintain awareness of patents that might cover the standard, and disclose them on an ongoing basis if the rules so require. This can be a difficult task for firms with large patent portfolios and inadequate management oversight, where no single person or department possesses overall knowledge of the company's patent assets or what they individually cover. A company lacking institutional awareness of its IP coverage risks inadvertent violation of standards obligations. Even if unintentional, such conduct may result in sanctions and lawsuits – particularly where use of the standard requires payment of royalties.

In general, companies involved with standards should therefore consider the following strategies:

- Develop and maintain firm-wide knowledge of patent assets. Know what patents could possibly cover technology contributed to standards bodies.
- Obtain the standards organization's guidelines before participating, and carefully document all compliance efforts. Determine whether the organization will review proposed license terms.
- Investigate before adopting. Determine the costs of using an industry standard before beginning design efforts.

Now to the second question: should Mouthscape go to the trouble of patenting its imaging convention or simply publish it? The answer depends on whether control over the standard provides Mouthscape with some strategic benefit. It may well; without control, a standards originator can find itself watching ruefully as the industry distorts its contributions, perhaps even to the point that they are no longer compatible with the originator's own products. For years Microsoft Corp. and Sun Microsystems Inc. dueled over the Java programming language that Sun introduced in 1995. When Microsoft came out with its own version, Sun sued, claiming Microsoft's real agenda was to saddle its customers with a substandard implementation – one which, incidentally, was incompatible with Sun's own implementation – and thereby confuse developers, ultimately driving them away from Java and on to Microsoft's .Net platform. The final history of the "Java Wars" has yet to be written, but the lesson is clear: even a developer intent on giving technology away should try to retain control of it through IP, if only to prevent its disfigurement or sabotage at the hands of competitors.

The mature company

Derek ages, Mouthscape prospers, and the company has begun to diversify – its dental division expanding into more barbaric devices such as drills and extractors, and its 3D imaging division now serving customers outside the dental world. As these divisions expand and ramify further, what kinds of growing pains will Mouthscape encounter? And how should its IP management efforts change in response?

Responsibility for IP administration will naturally diffuse to managers with more direct departmental focus. That does not mean abandoning an enterprise-level strategy, however. Without strong, high-level oversight, dissimilar company divisions can drift into similar patterns. This is particularly so with respect to quantitatively measurable criteria. No division leader, for example, wants to be the laggard in terms of patents filed for or obtained, even if her division benefits little from patents. Widening numerical disparities provide a prominent target for criticism, and hence will be minimized, usually to the company's strategic detriment – unless upper management is serious about enforcing consistent standards rather than simply comparing results.

In particular, the most senior executives – and ideally the CEO – should give division leaders broad latitude to create IP strategies for their divisions, but insist on periodic progress reports grounded in financial performance. Strong divisions may forsake patent filings out of reluctance to commit the necessary technical resources, which are productively occupied trouncing the competition and developing the next generation of products. Success in the market is imperative, of course, but without IP protection, the next generation may be the company's last. Conversely, lagging divisions may overspend on IP as evidence of turnaround efforts. Yet even the strongest patents will not sell a single product. A sophisticated IP strategy cannot substitute for a marketing strategy.

Division-level progress reports can take the form of table 4.1, and top management would be wise to focus on the last two columns. It must not be afraid to ask the pointed questions: When will this product actually come into existence? Hasn't that product reached the end of its market life and, if so, why are we still pursuing worldwide patent protection? Senior company executives must have the cooperation of division IP managers to discriminate among the IP that is directly relevant to products in the current pipeline, IP relevant to future products or strategies within specified (and verifiable) time frames, and IP relevant to nothing in particular.

Licensing options should be considered for any IP associated with a strong division, and especially IP not presently associated with specific products. If a division is hot, it has obviously tapped into the marketplace zeitgeist, so even unused IP may be of interest to competitors or complementary businesses. But identifying licensing opportunities can be challenging; how to determine whether a competitor that is not presently infringing would welcome the prospect of doing so for a fee? One common expedient is to identify licensable patents and determine, electronically, which other companies' patents cite them. The problem there is that every company's patents lag its innovation by the number of years (usually at least three) it takes for a patent to issue. Patents issuing today reflect yesterday's priorities.

Another approach is to package related patents together in an attempt to define the technology they cover as an industry standard. That's fine if broad licensing is desired, but removes the possibility of an exclusive deal with a powerful player. There really is no formula for identifying license opportunities outside friendly (but wary) ongoing dialog with competitors, suppliers, and customers, as well as acute focus on industry transactions as they occur. Learning of a deal involving patents similar to those you would like to shop provides a perfect opening to approach the proud new licensee – maybe it would like to expand its brood.

Automated patent analysis

The difficulty of overseeing a growing IP estate branching in different directions, as well as identifying potential patent landmines and opportunities, has led to the rise of consulting services and software tools that attempt to automate the process. Often the product they provide is highly visual in nature, and can seem like a beacon of light through an otherwise opaquely confusing thicket. Unfortunately, that visual clarity can be misleading. Automated analysis is a dull-edged instrument, which, as it hacks through the thicket, can snag hunks of irrelevant material while chopping away the important bits.

Supplemented by business-specific analysis and an aversion to broad generalization, as well as an appreciation for their limitations, automated IP reports can often furnish useful data – and sometimes almost nothing of value. The following are some typical reports, what they purport to provide, and what they actually deliver.

- Patent "hit counts" attempt to track the numbers of patents issued annually for a particular technology. The idea is to see whether that technology is heating up or cooling off over time, and to identify the major players. The problem is that varying technical vocabularies compromise the ability to identify all relevant patents, and "patent lag" means any information is inevitably years out of date.
- Patent landscapes, the quintessential visual product of automated IP analysis, attempt to depict – in map-like form – how patents cluster around various technology areas. The idea is to reveal where competitors' research is most active, and where your own research might best be directed (away from the cluttered areas and toward the sparse ones). While gross numbers of patents sometimes suggest the competitive threat, patent activity need not correlate with research success or market traction. Some companies file lots of patents regardless of value, while even a single broad, pioneer patent in a new area – covering, say, a class of pharmaceuticals – can render the sparsest terrain quite inhospitable to newcomers. Landscape reports may have greatest relevance when employed comparatively; companies with similar patent distributions may be attractive candidates for merger, licensing, or collaboration.
- Prior-art citations in competitor patents provide fodder for numerous reports. The median age of citations in a given technology area is sometimes offered as a gauge of whether a technology is emerging and hot or mature and unexciting. That is a potentially misleading generalization, since older

citations may represent no more than the defunct prelude to a still-budding field, and in any case patent lag precludes knowledge of what is hot *today*. Analyzing competitor citations of your own patents can reveal licensing opportunities or willing acquirers of unprofitable business lines, but instead may simply identify your known competitors. Some reports show the patents cited most frequently by others within a given technology area, with the goal of revealing the most potent threats. But a patent's citation frequency often says more about its age than its importance, and even a much-cited patent may have narrow claims; patents are cited for their information value – that is, the breadth of their disclosures, not their claims.

- Inventor reports can show who is innovative within a technology segment and which companies obtained the benefits of their intellectual labors. It can also reveal who is egotistical enough to patent his every whim, or whose departmental managers are misusing patents to augment the visibility of their business units.

Automated IP analysis can produce vast quantities of data with ease – too much ease. The danger lies in drawing generic conclusions from what is really just raw data; numbers are seldom meaningful in themselves. Automated reports can provide a rough initial cut preceding further analysis and exploration, but it is imperative to avoid over-relying on breathless multicolor presentations – no matter how much they promise or how expensive they are. The cost of over-reliance is invariably far greater.

Commenting on a painter who had declared himself the world's best, the critic Clement Greenberg quipped, "He's not devoid of gifts, but he's minor." The same goes for automated IP analysis.

5 Surviving IP disputes

Poor Derek. Commercial reality being what it is, at some point Mouthscape will receive a letter accusing it of patent infringement or find itself sending such a letter. It is the price of playing in the marketplace. Although only about 1% of United States patents ever make their way into court, the sheer number in force at any given time and their importance to their owners make it likely that commercial technology players will, eventually, become embroiled in a patent dispute.[1] Mouthscape's initial reaction, either to a threat or to the opportunity to threaten, may well dictate the course of the ensuing dispute by abandoning or preserving options – perhaps even pre-ordaining the outcome.

The receiving end of a threat

A corporation may be a legal artifice but it is run by people and, when menaced, will react in human terms – often passing through the stages psychologists associate with grief or trauma: fear, denial, anger, guilt, bargaining, depression, and acceptance. No self-respecting company would admit to such emotional, unbusinesslike behavior, of course, and therein lies perhaps the greatest danger. In the throes of a deeply human response it is unwilling to admit to, Mouthscape, having been threatened by a large competitor at a critical business juncture (is there ever an opportune time to be accused of infringement?), may react in numerous self-defeating ways. And, once taken, a rash action cannot be revoked. Here is a hit parade of the most hazardous missteps.

Place head in sand. That is fear and denial – maybe if I pretend it does not exist, it will go away. It won't. Anyone taking the trouble to analyze Mouthscape's product and set the legal machinery into motion, accepting the risk of a

[1] Statistically, the amount of patent litigation in the United States grew substantially between 1985 and 2000, but the rate of litigation relative to the number of issued patents remained constant.

counter-attack, is unlikely to disappear. Moreover, by sending the demand, the patent owner has put Mouthscape on formal notice. Now Mouthscape must take reasonable steps to ensure it is not infringing the patent or risk a finding of "willfulness" – a censure that may seem more appropriate to the principal's office but which, in fact, carries a heavy legal penalty: up to a *tripling* of damages plus, in extraordinary cases, payment of the patent owner's attorneys' fees. (Whereas in many European countries the loser always pays for all or a portion of the winner's attorneys' fees, the "American rule," which many criticize as an incentive to litigation, presumes that, absent egregious conduct, each side pays its own lawyers.[2])

Express yourself. Anger may motivate Mouthscape to fire off a reply to the patent owner. How dare he accuse us of patent infringement! The typical reaction to a competitor's assertion of patent rights is that the patent is total garbage – how could anyone patent *that*, for crying out loud, everyone does it and has for years! – and, in any case, it doesn't cover us. Somewhere a little voice may whisper the wisdom of calling the company's patent lawyer, but managers deafened by passion find the issues quite straightforward and see no need for expensive and cautious lawyers who will delay the giddy satisfaction of a stern riposte. It is here that anger mingles with parsimony to dangerous ends. An irate Derek may be tempted not only to scoff at the patent owner's assertions, but to impart some justification for his scoffing. Such justifications, uninformed by knowledgeable analysis, can furnish the patent owner with unintended (but potentially lethal) admissions. Saying, "We don't infringe your miserable patent because we do it this way . . ." tells Mouthscape's opponent everything she needs to know if the patent does, appearances notwithstanding, cover "this way."

A reply in anger can also obscure the genuine merits of your case. Rational patent owners, even those with the temerity to accuse you of infringement, have no desire to litigate losing cases. A constructive opening – stating, for example, that company policy is always to respect the valid IP rights of third parties – followed by a sober analysis of the patent claims or the prior art may prove disarming. An angry response, by contrast, communicates stonewalling and practically invites a lawsuit. Other situations call for different tactics – a request for further information, an assertion of your own patent, maybe even

[2] In Germany, for example, the reimbursible cost is set by the court according to the economic stakes involved in the case. While the threat of paying both sides' costs furnishes a strong disincentive to litigation, those costs can be dramatically lower than in the United States. A full year of litigation in Germany may cost less than a month in the United States.

a pre-emptive lawsuit (more on that below). The optimal approach is never discovered through rage.

Be your own lawyer. If slightly more circumspect but still cheap, the angry threat recipient may have its engineers analyze the patent first and provide an assessment. Not only are the engineers unqualified to make the assessment – a patent is a legal document, not a scientific paper – but may well share the company's sense of outrage or feel an obligation to tell management what it wants to hear. Neither management nor the technical staff has the legal background or objectivity to analyze the merits of the claim. Lacking such analysis, Mouthscape cannot avail itself of mitigating strategies, such as avoiding the patent or conducting a validity study.

An internal analysis may stoke fear rather than anger, followed by resigned acceptance. Feeling defeated, Mouthscape may instruct its engineers to find a way, any way, around the patent. Precipitous defensive action, however, can sometimes be worse than an ill-considered offense. Designing around a patent unnecessarily, for example, may compromise a company's position in a competitive marketplace – all for nothing if the patent is not infringed in the first place. Certainly any recipient of a threat should perform a preliminary internal analysis, but to assist rather than replace the efforts of its lawyers.

But any internal analysis *should not be in writing* or leave a written trail – whether in the form of e-mails, correspondence, memos, or even handwritten notes. In the United States, all of these are subject to discovery by an opponent in litigation, meaning they can and will be used against Mouthscape in court. A limited exception protects communications with the company's attorney, but there are exceptions to the exception. Now, it is far easier to recommend a "nothing written" policy than to carry it out. A company faced with a claim of infringement must communicate, and the need for effective communication may sometimes override precautionary practices. Know the risks and decide.

Tell your friends. The need for communication does not extend beyond the company doors, however. Customers may ask Derek how he plans to handle that patent claim. Nosy reporters may invite him to tell his side of the story. Friends may express sympathy. The temptation to talk – to run some exculpations up the flagpole and hope for approving salutes – can be terrific, surprising even the talker.

It is a temptation that must be stoutly resisted. Derek cannot assume strict discretion on the part of his interlocutors; they talk, too, and guilty extenuations can easily metamorphose into admissions. Disparaging your opponent can lead to a slander lawsuit or a claim of unfair competition. There will be plenty of time to put a public face on the dispute – after you have put a strategy together.

Eat the evidence. What happens if Derek remembers the rule against reducing thoughts to writing after having written? Should he delete those compromising e-mails? If they existed before Mouthscape received the demand, then their destruction can amount to tampering with evidence, leading to sanctions and/or the court's presumption that what was destroyed favored the accuser.

Be careful, in this regard, even when attending to the company's routine file-destruction policy. Do not adopt that policy, for example, after you receive a demand. Do not conspicuously "remind" employees about the policy at that point, either. In December 2000 a wealthy investment banker, who was responsible for the Wall Street debut of numerous big-name technology companies, encouraged subordinates to "clean up" their files while a federal probe loomed. Four years later, based on that 22-word e-mail, he was convicted of obstruction and witness tampering and sentenced to one and a half years in federal prison.

Self-serving as this may seem coming from a lawyer, the only right way to react to a letter accusing you of infringement is to call your patent attorney and formulate a plan. That plan will probably begin with an infringement analysis – a careful review of the patent, its claims, and its prosecution history. In the United States, the record of negotiations between a patent applicant and the patent office can strongly influence the way claims are interpreted. Similarly, claim terms can be limited or even altered in meaning by how they are used or defined in the patent text.[3] What superficially appears a formidable threat may, upon closer analysis, turn out to be harmless. Alternatively, the analysis may stimulate a discussion of design alternatives. An earnest effort to avoid patent claims, even if ultimately unsuccessful, can at least preclude a finding of willful infringement.

If infringement seems likely and no design-around strategies present themselves, a validity study may be the best alternative short of settlement. Although patents are presumed to be valid (more on that soon), they can be overturned

[3] Lewis Carroll seemed to have patent lawyers in mind when he wrote:

"When *I* use a word," Humpty Dumpty said, in rather a scornful tone, "it means just what I choose it to mean – nothing more nor less."

"The question is," said Alice, "whether you *can* make words mean so many different things."

"The question is," said Humpty Dumpty, "which is to be master – that's all."

Like Humpty Dumpty, the patent drafter is master of the words she uses and how she defines them.

by showing that the claimed subject matter was known prior to filing or, in the United States, before its invention or more than one year prior to filing. The patent office, of course, already has performed a search and granted the patent despite what it found. But the examiner might have missed something, or may simply not have had access to a journal article or conference paper. Locating that lethal needle in the vast worldwide haystack of relevant literature can require substantial investment; specialized searching companies have built lucrative businesses around their expertise at scouring obscure information repositories. A German-language abstract published in Prague and available in a single musty library is just as potent against a United States patent as an award-winning article from *Science*.

Of course there may, in the end, be no needle. Or maybe none sharp enough to lance all of the claims, each of which must be considered separately. Narrower claims, remember, are more difficult to invalidate. Independent of search costs, validity studies tend to be quite expensive – $10,000 is usually the floor for even a single patent – because each of the claims must be analyzed in excruciating detail based on, once again, the patent itself, its prosecution history, and the common understanding among those proficient in the technology.

To sue or not to sue?

What if Mouthscape obtains patent coverage for its prosthesis system and discovers a competitor selling an infringing product? What strategic considerations inform the decision to file suit?

In chapter 2 we reviewed the basic progress of a lawsuit in the United States, but our primary focus was the discovery process. Here we will consider the factors for and against litigation as a mechanism to resolve disputes, then some alternatives.

Any company even considering a lawsuit must perform some internal soul-searching. Part of that process is simple due diligence: know the realistic scope of your patents, identify potential weaknesses, develop as much information about the product (or process) you suspect infringes, and identify the specific goals you hope to accomplish through litigation. During the runup to trial, a court will dissect the patent claims and decide what they mean. Before deciding whether to sue, then, undertake a thoughtful analysis of how that is likely to turn out. After key terms have been defined and the scope of the claims bounded, will they still cover the competitor's product? For example, does your patent really teach what the competitor sells, or is it necessary to stretch the claim language or interpret it very literally in order to cover the competitive product? You may still win, but the odds decrease with each deviation from the patent's central teaching.[4] At the extreme edge of the interpretive spectrum lies the doctrine of equivalents,[5] that is, the hope that a court will expand the claims to embrace more than they literally cover. It is rarely advisable to file based on that hope alone.

Determining the fit between your claims and the product at issue depends just as strongly on your knowledge of that product. Can you get your hands on a sample and have it characterized by a neutral expert? Does the manufacturer claim conformance to a standard in which you have patent rights? If so, does your patent really "cover" the standard in the sense that every possible implementation necessarily infringes? Sometimes an uncooperative competitor makes it necessary to file suit just to get enough information to assess infringement. But, just as often, cooperation may be elicited. Keep your inquiry focused and specific – you need just enough information to verify your competitor's denials of infringement. Of course, you may also need more than self-serving (and possibly misleading) explanations. If a process is involved, you might offer to have a mutually acceptable laboratory review it for relevance to the patent; the laboratory can issue its report directly to your attorney, for her eyes only, to protect trade secrets. A similar approach can be used for pre-release products or those that involve proprietary formulas or hidden characteristics. Explain that the goal is to avoid litigation which, if those protestations of non-infringement are to be believed, your competitor would win anyway – after needless expense on both sides.

Finish the due diligence by examining your patent for obvious weaknesses:

- Have any potentially damaging prior references been located since the patent issued? If so, you may want to consider reissuing or re-examining the patent through the patent office before litigating it. (If they were known to you during prosecution and you failed to disclose them to the patent office, you're in trouble – particularly in the United States – and you'd *better* have the patent re-examined.)
- If you did not perform a patentability search before you filed for the patent, perform it now; your opponent certainly will. Even if you did, consider

[4] Patent claims typically cover more than the specific examples provided in the patent text, and, in the United States, case law precludes courts from limiting patent claims merely to what's disclosed in the patent. At the same time, claim coverage must ultimately be commensurate with the patent's teaching. As a matter of fairness, therefore, courts may also refuse to let a claim cover subject matter which, while literally covered by the words, departs so significantly from the patent's teaching that coverage is largely accidental. Drawing the line between fair and excessive claim readings can be a Solomonic exercise.

[5] See chapter 2.

commissioning a more intensive one. Better to learn the weaknesses before you file suit, while you can do something about them. Once again, the patent can always be reissued or re-examined.

- Review inventorship. In the United States, inventorship is a purely legal determination, and, if the patent lists too many or too few people – for example, naming a marginal contributor as an honorific, or omitting an inventor because he left the company before the patent application was filed – the patent is *invalid.* Inventorship mistakes can often be cleaned up, but not if deliberate deception was involved.
- Consider having a law firm other than the one that drafted your patent review it and the competitor's product, and perform an infringement study. You want to discover all the warts on your case.

In the United States, patent owners tend to fret more over infringement questions than the vulnerability of their patents to attack. This is because patents enjoy a legal "presumption of validity": the law presumes that the patent office knew what it was doing when it issued the patent, and casts a heavy legal burden on anyone challenging its validity.[6] It is very difficult, for example, to overturn a patent by alleging that the examiner failed to give enough weight to this or that prior reference; the court will usually defer to the examiner's judgment. The situation is dramatically different in other countries, where courts happily second-guess the patent office. Still, deference notwithstanding, United States courts find only around 30% of litigated patents both valid and infringed; the remainder provide no monetary award or other permanent remedy to their owners.

Assuming Mouthscape remains satisfied with the strength of its case (and the sobering statistics), it is time for some hard questions. What, precisely, do they seek to achieve through a lawsuit? Given the expense involved, the most obvious attraction hopefully is money – compensation for past damages and elimination of a direct competitive threat, which will enhance Mouthscape's market position and future sales revenues. Derek must quantify these factors to ensure that the spoils are worth the cost of battle.

Mouthscape may also see litigation as an avenue to licensing revenue. If war is a headbanging version of diplomacy, litigation can be a desperate form of negotiation – sometimes a small player must roar in order to be heard. This is particularly true for licensing start-ups. Unlike Mouthscape, which actually

[6] The law requires "clear and convincing" evidence of patent invalidity. In most civil litigation, a party must carry its burden of proof merely by a "preponderance of the evidence." To get a feel for the difference, think of the former as an abiding conviction, while the latter just means more likely than not. Both standards fall far short of the "beyond a reasonable doubt" benchmark for criminal convictions.

sells products into the marketplace and may have earned the grudging respect of large competitors, licensing newcomers may seem like parvenu shakedown artists merely out for a buck. Maybe they will go away, thinks the large company, they cannot possibly afford top-tier lawyers nor can their patents be all that great. Maybe they will run out of money and we will buy their technology on the cheap. These sorts of reactions stem less from disrespect of IP rights than natural resentment at what would amount to a tax on the competitor's own technology, which it has already spent great sums to develop. Unfortunately, that technology may tread on the rights of others who got there first, and lawsuits have been known to work wonders on attitudes – what at first seems like an affront may, following summary judgment, strike the loser as a fabulous business opportunity.

Licensors often have a contractual obligation to sue. No licensee can tolerate infringement of licensed IP; free-rider infringers, who do not pay the tax that the licensee has agreed to pay, will simply undercut the licensee's prices. They must be stopped or the license will have no value to either party. Which side has the right or the obligation to bring suit can vary. An exclusive licensee will usually bring suit itself, while nonexclusive licensees typically look to the licensor. A license may contain triggers (such as a measurable effect on the licensee's sales or the licensor's royalties) that obligate one party or the other to sue, and impose penalties (for example, loss of exclusivity or termination of the license altogether) if the obliged party fails to do its duty.

Less directly, Mouthscape may fear the results of *not* suing, particularly if the infringement is blatant. Passivity in the face of an IP violation all too often invites further violations.

Those are the positive strategic factors favoring an infringement lawsuit. Disadvantages abound as well. The disruption. The cost. The risk to the IP and to market acceptance of the technology, the prospect of being thrown on the defensive by an unfair competition counterclaim. Good litigators understand these dangers and, whenever possible, seek favorable settlement terms rather than outright victory.

Before crossing the Rubicon of litigation, outline the worst possible settlement terms you could possibly live with, and then consider alternatives. Is there any chance for a strategic combination with the adversary? Better explore that possibility before taking legal action. IP litigation differs fundamentally from a commercial spat, where only money is at stake. As noted earlier, a threat to a company's IP represents a threat to its identity, and, once a lawsuit is filed, a merger or acquisition may be off the table for good. The filing itself may be seen as doing unforgivable damage just in drawing the originality of the defendant's technology into question, motivating it to shun even an economically attractive combination – either out of corporate pique or fear that the marketplace will interpret the combination as a capitulation, undermining the defendant's position in future IP disputes.

When approaching its competitor with allegations of infringement, Mouthscape must tread lightly to avoid inadvertently yielding the initiative. In the United States, a too-threatening demand letter triggers "declaratory judgment" jurisdiction, empowering the targeted competitor to take *Mouthscape* to court rather than the other way around. In a declaratory judgment case, in other words, the accused competitor strikes pre-emptively, challenging the basis for the accusation and asking the court to "declare" the asserted patent invalid and/or not infringed. Filing first confers a host of procedural advantages, most notably the choice of timing and courthouse. It is critical, therefore, for Mouthscape to keep the temperature low as it researches infringement and explores alternatives to litigation – attacking, if it must, on its own terms and timetable.

Alternatives to litigation

Not every dispute winds up in litigation. Anyone wishing to challenge a patent can do so administratively, through the patent office, rather than in court. Procedures vary among countries, but virtually all permit the public to weigh in on the propriety of granting, or having granted, a patent.

Parties to an IP dispute may also opt to forgo the expense and disruption of litigation in favor of "alternative dispute resolution" (ADR), most notably arbitration or mediation. These procedures streamline the steps involved in resolving or settling conflicts, and give the parties varying degrees of control over the process itself. The complexity and (hence) cost of IP disputes makes them ideal candidates for ADR, in the view of some, while others hold precisely the opposite view.

In this section we will explore the factors for and against these approaches as alternatives to litigation.

Mediation and arbitration

The essence of mediation is structured negotiation. Two sides unable to speak civilly or view their dispute objectively at least agree on a procedure for talking, with the assistance of a mediator trained in ADR. For mediation to succeed, the goal must be settlement rather than triumph. The mediator attempts to help each side recognize the strengths and weaknesses of its position, and facilitate agreement based on mutual, if reluctant, acceptance of these realities. To irascible courtroom lawyers, the notion of mediation conjures images of stuffed couches, New Age music, and some scruffy social-service longhair uttering platitudes and leading a chorus of Kumbaya. Mediation offers no discovery, no referee with disciplinary authority over an uncooperative party, no opportunity to interview witnesses under oath, and no prospect for interim relief or an injunction. No one, in short, is really in charge.

Proponents of mediation view the absence of an adversarial environment in positive terms rather than as a downside. Shorn of the formality and stress of litigation, the mediation procedure gives both sides ownership of an agreed-upon outcome. The procedure is flexible, private – unlike a lawsuit, which produces reams of public records – and fast. As a result, of all dispute-resolution procedures, mediation provides the greatest opportunity to preserve relationships going forward.

Mediation, therefore, is best considered for disputes in which the underlying facts are simple (no discovery needed) and uncontroversial, and a relationship worth maintaining exists. Parties to a license agreement, for example, may disagree over which products fall within the royalty base, or how to calculate net sales, but otherwise view the business arrangement as productive. A mediator familiar with the relevant law can help the parties bridge the legal and financial gaps separating their positions – if the parties are open to compromise. Mediation should not be undertaken with inflexible positions or the objective of winning all the marbles. The mediator's role is to help both sides divide the marbles.

Mediation can be subject to misuse. People have exploited the flexibility of the procedure to stall for time, and power imbalances can undermine efforts at conciliation even if both sides profess open minds. There is no decision delivered in mediation or obvious final stopping point, so it is important to know when to call it quits.

On the spectrum of formality, available remedies, and authority over the parties, arbitration falls, predictably enough, between mediation and courtroom litigation. The parties generally agree on ground rules – the guidelines published by the International Chamber of Commerce are popular – and then select one or more arbitrators according to those rules. Often each party picks one arbitrator (subject to the other side's objection based on conflicts of interest and the like), and the two chosen arbitrators appoint a third. The ground rules can provide for limited discovery, injunctive and interim remedies, and a degree of oversight that typically depends more on the personality of the arbitrator(s) than the rules of engagement. And, unlike mediation, neither side is free unilaterally to terminate the proceeding if it dislikes how matters are progressing. Usually the arbitration culminates in a hearing at which each side presents its case, perhaps interviewing witnesses (though not under oath). The arbitrator's decision, once rendered, is binding on both parties. Courts may intercede to enforce that decision, but absent outright fraud or an out-of-control arbitrator, will rarely disturb it if one side is unhappy.

The primary advantages of arbitration are speed and low cost, privacy, and the ability to select arbitrators for their expertise in a particular technology or business; litigants, by contrast, are stuck with the judge they draw. Complex disputes involving difficult technologies, numerous players, and the need to preserve trade secrets can theoretically be accommodated through agreed-upon discovery procedures. But what happens if one side stonewalls, or is less than fully forthcoming? It may be just as difficult to compel cooperation as it is to pull the plug on the arbitration altogether, resulting in an inexorable march to a decision based on a less-than-adequate record.

The stereotypical view of arbitration is a fast-paced proceeding in which both sides present elaborate cases to an arbitrator who will simply split the difference. Like all stereotypes, this view is unfair but revealing. Arbitrators do tend to view themselves as engaged in a process of resolving, rather than deciding, disputes.[7] And, like mediation, arbitration is best suited to controversies involving clear factual issues with parameters upon which both parties agree – the amount of money due under a license, whether a product conforms to industry norms or contract specifications, or a patent dispute involving a publicly available product and undisputed claims. As legal questions multiply or if discovery grows complex, arbitration's cost and speed benefits over litigation diminish, and its disadvantages, mainly the lack of strong central control and institutionalized authority, become pronounced.

Table 5.1 summarizes, in necessarily oversimplified form, some advantages and disadvantages of the different ways a dispute can be resolved.

Re-examination and opposition

The examiner's willingness to grant a patent does not mean it is over. Most countries provide some form of pre-grant or post-grant opposition that allows

[7] It is important to stress that everything, including the arbitrator's discretion to make compromise awards, is within the discretion of the parties. They set the rules. "Baseball-style" arbitration, for example, requires the arbitrator to choose between each side's proposed terms of decision without modification.

Table 5.1 Dispute-resolution alternatives

	Mediation	Arbitration	Litigation
Adversarial	No	Somewhat	Yes
Appeal available?	No	Rarely	Yes
Confidential	Yes	Yes	No
Cost	Lowest	Somewhere in between	Highest
Who's in control?	Parties	Arbitrator (subject to agreed-upon ground rules)	Judge
Discovery	No	Limited	Yes
Injunctions	No	Sometimes	Yes
Speed	Fastest	Somewhere in between	Slowest
Stress	Low	Some	Plenty
Witnesses	No	May testify	Testify under oath

outside parties, or even the applicant herself, to ask the patent office to consider (or reconsider) issues relating to patentability. Oppositions must be commenced within a fixed time window, usually nine months following patent grant or a similar period after the examiner publishes an intention to grant. After the opposition period expires, the patent can no longer be challenged through the patent office; an opponent's only recourse lies in court. The US re-examination system, by contrast, permits outside parties to challenge a patent at any time during its life.

Oppositions have particular importance in Europe due to the nature of European patent practice. As explained earlier,[8] a European patent is not itself enforceable; rather, it must be validated – that is, translated and registered – in European countries where enforceability is desired. The European patent, in other words, fragments into a series of "national" patents, each with its own life. So a successful court challenge to, say, the resulting French patent will have no direct effect on the German or Italian patents. The courts in different European countries have varying procedures, and may well reach inconsistent conclusions concerning the patent's validity and scope. Only the opposition mechanism allows for a single, centralized legal challenge. The outcome of the opposition, which can range from outright cancellation of the patent to narrowing of its coverage to no change at all, will be propagated to every European country in which the patent is validated.

All prerequisites to patentability – novelty, inventiveness, adequacy of the patent's teaching – are fair game in an opposition. The opponent need only

[8] See Chapter 1.

lay out his case before the European Patent Office, and may participate in the proceedings as they inch along. The patent owner or the opposer may appeal the outcome of the opposition. On average, an opposition (including appeal) takes more than three years to resolve, further delaying patent enforceability beyond the four or so years that typically elapse before a European patent is granted in the first place. These time periods subtract from the patent's life, which is measured from filing. Given the relatively low costs[9] involved, the relatively high chance of success – about a third of all opposed patents are revoked entirely – and the prospect of substantially delaying patent enforceability in any case, oppositions have proven quite popular; nearly 8% of all European patents encounter opposition.

Faced with persistent complaints of a system stacked against patent owners, Japan recently scrapped its opposition system in favor of "invalidation trials," which resemble the US re-examination system, before the Japanese Patent Office. In the United States anyone, including the patent owner, may request *ex parte* re-examination of an issued patent. If the requester can convince the patent office that a substantial new question of patentability exists, prosecution will be reopened, and the patent owner must confront the new issues. As in an opposition proceeding, a patent may emerge unscathed, with narrowed claims, or not at all. But the only eligible questions involve novelty and inventiveness; the sufficiency of the patent itself, open to attack under the European opposition system, remains off-limits.

Once the request is granted, the requester largely drops out of the picture unless he is also the patent owner. And nearly half of all re-examination requests are, in fact, submitted by the owners themselves. They come seeking absolution for past sins (failure to cite an important prior reference, usually) or a benediction for the future (in the form of claims confirmed over a newly discovered piece of prior literature). Remember that presumption of validity: difficult as it is to attack a patent based on a reference considered during prosecution, it is even harder when the reference was the focus of re-examination.

Third parties, by contrast, rarely initiate *ex parte* re-examination of patents they wish did not exist. Whereas opposition proceedings offer low strategic and financial cost, essentially the opposite is true of re-examination. After the requester submits his argument and convinces the US PTO to re-examine the patent, the action shifts to the patent owner and the examiner – often the same examiner who allowed the patent in the first place. The original examiner, then, is called upon to second-guess his own earlier handiwork.

[9] Generally €15,000 to €25,000 from start to finish.

Although the requester is kept informed of developments, he must remain on the sidelines, saying nothing as the patent owner submits her arguments, perhaps even presenting them in person to the examiner, and tweaks the claims. The requester, in other words, can do no more than toss a hand grenade into the room, which the patent owner and the patent examiner proceed to defuse. It is no wonder that only around 12% of *ex parte* re-examinations result in cancellation of the patent. Better to test validity in court, most opponents of a patent conclude, where both sides can present arguments.

Widespread dissatisfaction with the asymmetry of *ex parte* re-examination led the US PTO to institute a new system of "inter partes" re-examination. In this proceeding, the requester has the right to rebut arguments made by the patent owner, and in-person interviews with the examiner are prohibited; that means the entire proceeding is conducted by paper submissions. Either side may appeal the decision. The average cost is about $100,000 per side, and the patent office's goal is to resolve inter partes re-examinations within 21 months (although many cases have taken and will take longer).

From the perspective of someone seeking to avoid a patent, inter partes re-examination represents a clear improvement over the *ex parte* system. While the time frame may be comparable to litigation, the costs are dramatically lower, and procedurally both sides share a level playing field. Moreover, the presumption of validity that helps a patent owner in court does not apply in re-examination.

Still, there may be reasons for a would-be requester to be wary. Only novelty and inventiveness will be considered, and only with reference to printed publications. In court, by contrast, a patent can be attacked on the basis of earlier commercial product sales, public disclosures, inequitable conduct before the patent office, or some insufficiency (inadequate teaching or claim vagueness, for example) in the patent itself.[10] Moreover, in a re-examination proceeding, the patent owner can always narrow her claims. In litigation the patent claims are frozen in their issued form, either surviving the challenge or shattering completely. Patent lawyers always worry that, with the opportunity for strategic retreat through amendment and argument, the patent owner will probably wind up with *some* coverage when the re-examination concludes, and that coverage – now largely bullet-proof against prior literature, having withstood the best that the requester could hurl at it – may well cover the products his client wishes to introduce.

[10] Of course, nothing prevents a requester from raising these issues in litigation following the re-examination. What he cannot do is relitigate the validity of any of the re-examined claims in court, at least with respect to prior art that was "available" to him during re-examination.

Related to re-examination, although by no means an alternative to litigation, is "reissue." In the United States, a patent owner can ask the US PTO to reconsider his patent and reissue it with new or changed material – claims, text, drawings, the list of inventors, you name it. The purpose of the reissue process is to allow patent owners to correct mistakes that compromise enforceability; a patent naming the wrong inventors, as noted earlier, is invalid. The patent office interprets the notion of a "mistake" liberally, however, and, if filed within two years of the patent's issuance, a reissue application can request expanded claim coverage – a feat impossible in re-examination. Reissue can be a complicated business (the patent office seems to hate resuscitating the dearly departed through reissue), and like re-examination, it places the entire patent at risk. You may ask the patent office to reissue only some claims, but the examiner will review all of them, and you may wind up with less than what you started with.

IP litigation insurance

Insurers have recently taken to writing policies against IP litigation. While most are defensive in nature, covering claims made by IP owner against the insured, some cover offensive activities – funding litigation, for example, against infringers of the insured's patents. The terms of such policies vary wildly and from year to year, as carriers learn painful lessons or search in vain for a market. With premiums and deductibles both typically high, businesses have not exactly signed up in droves for IP insurance, but it may serve quite well in specific circumstances – for example, if a small company sees IP disputes as all but inevitable, IP insurance may allow it to sustain otherwise unaffordable litigation over the long haul.

6 Giving diligence its due

Conducting an IP due diligence means investigating and evaluating a company's IP assets, practices, and risks. How strong are the patents? Will they help the company succeed? Do outsiders – an inventor's former employer or university, competitors bristling with their own patents – have any potential claims? Most commonly, IP due diligence precedes a transaction such as a round of venture funding, a merger, or an acquisition. Not only do professional investors hate to watch their money fund lawsuits, but also must measure the worth of IP assets in order to place a realistic value on prospective investments. A company's market projections may ultimately rest on its ability to secure a strong IP position, and, moreover, that position can itself prove a valuable asset even if the company fails.

Too bad, then, that IP due diligence can often seem a ritual ticking off of checklist entries rather than an effective search for meaningful issues. The grand menu of items that might be considered is enormous – if you do not believe it, feast on the banquet that ends this chapter. But performing an IP due diligence should not feel like being lost in an endless, expensive buffet, and, if it does, it's time to find a different restaurant. Context and common sense dictate the proper tasks. If a company's patents do not figure strongly into its business strategy (or an investor's estimation of what that strategy should be), why spend time reviewing them? If its products have been on the market for years with little change and no legal challenges, does it really make sense to investigate third-party rights?

Perhaps it's fear of high cost and mechanical execution that has led to underutilization of this important forensic exercise. IP due diligence is not just for venture capitalists and M&A players. An IP-centric technology company should have the internal expertise to undertake its own periodic self-examination – particularly after reorganizations, when new management may be tempted to change nothing or too much. Similarly, outside candidates for executive positions should scrutinize the IP profiles of their prospective companies – the blemishes as well as the flattering features – before signing up.

And no one should license a patent before investigating its pedigree as well as the possibility that someone else out there has blocking rights.

Consumers of IP due diligence need an understanding of what they are getting and, more importantly, not getting. No amount of effort can identify all possible problems or erase them. It is important, rather, to think in terms of minimally acceptable comfort levels. It is also useful to distinguish between inward-looking due diligence, which explores the quality of a company's own IP, and outward-looking efforts that consider defensive risk. The distinction is imperfect and spillover inevitable, but beginning a project assessment by deciding which is more important can help focus broad priorities at the outset, avoiding premature descent into checklist-level detail.

CASE STUDY #4: Sandy Pope, a principal in the venture capital firm of Hope Springs Eternal LP (HSELP), is trying to make sense of a business plan. The company is some outfit called True Blue, Inc. and they hope to corner the world market for blue lasers. The risk factors all look great – huge customer base, high margins, working prototype developed in less than a year, no professional investment yet. On the other hand, their competition includes immense industrial powerhouses, and, as yet, True Blue has no patent protection. It isn't clear if they even have an IP strategy, despite all the R&D work. Very odd. Maybe it's just a case of inexperience, yet they've raised nearly a million dollars in "angel" funding and even amateur investors don't usually plunk down so much cash on an unprotected idea.

Sandy knows something about semiconductor manufacture, and the business plan's description of the facility mystifies him further. These guys have a full-featured clean room, for crying out loud, Sandy thinks. It must have cost millions. Where did True Blue get all that expensive equipment? Certainly not from its angel investors, if the financials are to be believed; their invested capital would barely cover the first year of operations at starvation salary levels. And that Ovkorsky guy. Something secretive about him. Maybe Sandy has read too many Cold War novels, but he wonders if Dmitri is holding anything back.

Still, the deal terms are too good to ignore, misgivings aside. HSELP's limited partners will plant their pelf in savvier soil if True Blue hits big after Sandy turns them down.

What should Sandy investigate?

What's the strategy?

Almost every IP due diligence begins with an inventory of the target company's IP assets. A status list of all current patent and trademark applications, as well as disclosures being evaluated for patent potential, should be requested immediately. While Sandy is waiting, though, he ought to consider exactly how patents will fit into True Blue's business strategy. It is clear they will be central:

True Blue is an innovation enterprise, one whose market prospects rest solely on its proprietary advantages. Whereas some companies may hope to succeed based on service or consulting capabilities, price competition in a commodity market, or custom product tailoring in a differentiated market, True Blue's fortunes are hitched solely to the demand-fulfilling value of its innovations – a value that will disappear the moment its heavy-hitting competitors gain uncontrolled access to them. Put differently, a True Blue investor's ability to realize a return, whether by acquisition or public offering, will depend on the company's ability to prevent or ration competitive use of its technology. Sandy is thinking of investing not in True Blue's brains, but, ultimately, in its IP. And, in an IP play, inward-looking due diligence must be the first priority. At a minimum, Sandy should consider:

- The likelihood that True Blue will obtain worthwhile IP protection for its core innovations.
- Whether True Blue has taken, and has established routine practices for taking, steps to secure ownership in its IP.
- Whether that ownership is free of claims by outsiders.

Quality of protection

Let us say Sandy receives a status list along the lines of table 4.1, and that it contains half a dozen entries – all relating to process recipes. Dmitri and True Blue's CEO have convinced Sandy that the control software is best protected as a trade secret. Very nice, Sandy thinks, but what does the list tell me?

Not much, other than the level of True Blue's efforts and commitment to a patent strategy. The patent office has reviewed nothing at this point, so those applications represent no more than humble requests. True Blue's commitment, therefore, may represent anything from a solid foundation to a complete waste of resources to an expensive smokescreen – an attempt, that is, to project the appearance of IP value where none exists. A Potemkin patent village, Sandy muses. Could Dmitri be one of those Potemkin types?

The only way to find out is to perform a patentability search. We made brief mention of such searches earlier; they involve a relatively shallow but broad perusal of the prior art to locate the literature likely to be considered by the patent examiner, with the goal of predicting the degree of protection he will likely grant. A patentability search is like a biopsy – it should be thorough enough to locate obvious problems, but not so comprehensive (that is, expensive) as to reach the point of diminishing returns. Professional searchers know how to plumb the literature with ruthless efficiency. Often they confine their efforts to patent literature, however, in which case Sandy (or his lawyer) should supplement those results with a more focused search of databases covering relevant conference proceedings and academic journals, or at least a Google query. Searches can also be staged; perhaps a quick biopsy will tell Sandy everything he needs to know, and, if not, further effort may be undertaken. At the end of the search process, Sandy's patent lawyer can easily review the results and forecast whether True Blue's patenting efforts will succeed.

Unfortunately for Sandy's wallet, patentability searches are performed on a feature-by-feature basis. It really is not possible, in other words, to search at the "product" level if a product has multiple potentially patentable features. Unless those features are just different versions of the same new idea, rather than new ideas in themselves, they cry out for separate searches. Listen to their cries. A search for gasoline engines will turn up improvements to carburettors and valves, plus a few railroad cars full of unrelated junk. Do not cast too broad a net in an effort to economize; it just does not work.

On the other hand, perhaps Sandy can avoid this undertaking by leveraging the efforts True Blue has already made – or can be persuaded to make. True Blue should be obtaining patentability evaluations as a matter of routine. Now, it may be reluctant to share these with Sandy; a patentability analysis represents legal advice, and, as such, falls within the "attorney–client privilege." Created to encourage candid discussion between between lawyers and their clients, the privilege shields communications (such as opinions) from the prying eyes of an opponent in litigation. But the privilege can be waived if communications leak outside the attorney–client relationship. So, True Blue's CEO may protest, we cannot divulge those patentability analyses to you now without sacrificing the privilege, which may compromise us if we ever sue on our patents. To which Sandy may respond in any of three ways. Fine, he might say, just send us the references – they are public anyway – and we will evaluate them ourselves. Or he might offer a "common interest agreement," which attempts to extend the attorney–client privilege to HSELP by outlining the alignment of legal interests between it and True Blue. Or, finally, Sandy can simply insist, contending that the risks are minimal even if the privilege is lost. No competent lawyer writes a negative opinion, lest it leak out and potentially cause far more harm than its informational value could ever justify. (Imagine the courtroom histrionics: "Ladies and gentlemen of the jury, True Blue's own patent counsel cautioned them – nay, virtually pleaded with them! – not to file for this patent, knowing it could never withstand scrutiny . . .") And a positive opinion only helps. So cough it up.

Let us say that Sandy obtains the opinion. How does he evaluate it? Lawyers often go to great lengths to put the best face on a pathetic patentability case, particularly if they know their client's financial survival may depend on a favorable impression. A "clean" opinion tells a simple story, unburdened by mental gymnastics, of an invention that is different from prior work and why it is different enough. The more an opinion deviates from this platonic ideal, the less likely its rosy predictions are to come true. A common tactic, for example, is to attack the *quality* of an obviously relevant piece of prior art. While it is true that a reference must provide a sufficient teaching before it can dash the fond hopes of a patent applicant, the level of teaching required is actually quite modest. A reference can *invalidate* a patent claim, for example, based on far less teaching content than is required to *support* that claim in the application. Questioning the adequacy of a reference's teaching, like pounding one's shoe on the table, is often the last refuge of the desperate and the damned.

Sandy will also want to get his hands on True Blue's patent applications. Since these are confidential until they are published 18 months after the priority date, True Blue may ask HSELP to sign a nondisclosure agreement first (although professional investors, wary of limiting their freedom to consider related investment opportunities, often resist such agreements). Sandy should consider the quality of the applications according to the criteria outlined in chapter 4. But his most important focus as an outside investor is the degree to which the patent claims coincide with True Blue's business strategy. If, for example, the claims focus on specific process recipes – or if the claims are broad but the prior art is likely to restrict them to recipes – then True Blue may be in trouble. Sandy must look for any gaps between the market exclusivity True Blue needs to prosper and the patent coverage it seeks (and is likely to obtain). An easy analysis this is not, often requiring the assistance of an industry expert or technical consultant. For example, even a narrow patent can cover the universe if nothing outside the claims will work. How sensitive are True Blue's recipes? Does crystal growth diminish gradually outside the range covered by the patent applications, in which case the patents will only confer an incremental advantage (since non-infringing conditions may be good enough)? Or does the process fail entirely outside that range, offering the possibility of market-dominating IP protection? No analysis confined to the IP itself can answer these questions. It requires highly specialized expertise (or an in-depth discussion with, and a great deal of trust in, Dmitri).

Those who perform inward-looking due diligence generally seek to assess patent strength; they are often surprised to learn that without an industry perspective, diligence can only reveal weaknesses.

International protection must also be considered. If a company plans to address foreign markets, as True Blue does, at the very least it should have preserved foreign rights with PCT applications. But assessing quality goes beyond reservation of rights. Patent applications must be drafted based on the broadest protection available in any country in which the applications might be filed. Methods of therapy, for example, while mostly unpatentable in Europe and Japan, qualify in the United States. Therefore, a European application drafted for compliance solely with, say, European law will be inadequate for the United States if it contains insufficient support for therapy claims; such an application needlessly leaves the possibility of valuable US protection on the table.

Does True Blue have all the rights it needs?

Sandy and his lawyer have reviewed True Blue's patentability opinions and patent applications. The opinions are convincingly positive and the patent applications seem effective. HSELP's industry consultant confirms that the patent coverage sought, and likely to be obtained, will prove potent. One omission noted by the lawyer is the apparent absence of patent assignments from the inventors to True Blue. The files, and patent-office records, show nothing recorded for any of the applications. Just make sure it's done before closing, she says.

But Sandy is troubled. He knows that without a written assignment, True Blue's ownership of the patent applications is in doubt. He wonders if the omission is innocent. A colleague once told him about the president of a small company who thought he could keep key patents for himself and license the company later; naturally HSELP steered clear of that investment. Could Dmitri be pursuing a little self-dealing? Of course, if this were an espionage novel, the truth would involve more intrigue. Dmitri might be working for someone, some shadowy Mr. Big. Or someone could be blackmailing him. Someone from the old country – someone who knows something about Dmitri's past.

Foolish, idle thoughts, Sandy realizes. Must . . . stop . . . spinning B-movie plots. But still. Is there some way to allay suspicions, however unreasonable?

As explained in chapter 4, IP assignments are essential and must be obtained at the earliest possible stage. Employees and consultants move on, not always on pleasant terms, and a resentful former employee may, shall we say, interpret his residual obligations narrowly. Sandy should ask True Blue who, exactly, contributed to the development of the core technology. Hopefully the same people named as inventors on the patent applications, but it is always good practice to ask the question directly. Start-ups do not always understand how limited their discretion is in naming inventors, and patent lawyers do not always probe as searchingly as they should. The sooner discrepancies are recognized, the more likely it is they can be corrected.

In addition to invention assignments, Sandy should have his lawyer review True Blue's employment agreements and contracts with consultants. He should verify, first, that all individuals who have contributed to the company's research have actually signed an agreement. He should also assess whether it is sufficient. A solid contract cannot transform a resentful departing employee into a cheerful contributor, but at least it can restrict his legal options. A good agreement should contain:

- A nondisclosure obligation to refrain, both during and after employment, from divulging confidential business or technical information.
- An invention assignment clause that includes an obligation not only to assign rights, but also to promptly disclose the invention to the company (which cannot pursue what it does not know about) and to assist, even after termination of employment, in securing IP rights.
- An assurance that the engagement will not breach obligations to prior employers (for example, with respect to proprietary information or a covenant not to compete).
- A list of prior inventions that the company and the employee agree were made before employment.

True Blue should happily provide Sandy with copies of these agreements and whatever invention assignments exist (preferably after HSELP signs a nondisclosure agreement). For his part, Sandy should make sure everyone listed as an inventor on True Blue's patent applications has signed an agreement, and if not everyone has that at least they have executed invention assignments for the specific patent applications on which they are named. This exercise will help clarify the research staff's obligations to the company and the degree to which True Blue has secured its rights.

What it will not provide is an outward-looking sense of whether any of True Blue's employees owe a conflicting duty to someone else. While it is nice if everyone has signed an assurance against such conflicts, assurances will not satisfy the likes of Sandy. At least with respect to the core research personnel, Sandy may want to determine where they came from and when they began work for True Blue. Then he might investigate the possibility of conflict. This review begins with simple questions concerning past inventor affiliations, and should be supplemented with an electronic search for patents and applications naming those inventors – a worthwhile means of verification that can also provide a more detailed picture of researchers' past experience. It can also reveal potentially problematic patents owned by others.

Suppose, for example, that one of Dmitri's co-inventors recently earned her doctorate. Should Sandy worry if her thesis topic seems uncomfortably close to her work for True Blue? Universities typically have policies in place defining the obligations of graduate students and faculty. In the United States, most academic institutions minimally insist on owning all IP growing out of research conducted on campus, using university-owned facilities. But the obligation can be broader than that. While faculty may enjoy the prerogative to devote a certain percentage of their time to outside ventures, free of university commitments, graduate students rarely have that option. Some universities have adopted aggressive IP policies that claim ownership of everything students think up during their period of graduate servitude. Others are less draconian. In Canada and Europe, for example, institutional ownership is not automatic; many universities have ambiguous policies that share rights with the faculty. Whatever the policy, it probably covers students and faculty regardless of whether they have signed an agreement to that effect. So, if any True Blue inventors have had recent academic affiliation, Sandy should consider how much time passed between the end of that affiliation and the beginning of employment at True Blue, and how similar the academic research was to that performed for True Blue. Nearness in time and subject matter should prompt a review of the institution's policies, and may ultimately precipitate a request to the university for an ownership waiver or, more likely, a license.

Similar concerns apply to past employment. (*Maybe*, it occurs to Sandy, *Mr. Big is Dmitri's former employer! Dmitri knows he can't assign to True Blue what his old company owns, so he's holding back on True Blue, hoping he can cut a deal with the mentor he betrayed . . .*) Here the question is one both of contract – is there a written agreement with the former employer? – and law, which may supersede onerous contract terms. Courts attempt to balance the proprietary interests of employers with the right of workers to seek gainful employment. A business may legitimately restrain its former employees from divulging specific, identifiable trade secrets. But it cannot prevent their use of general knowledge and skill learned on the job. So True Blue is ordinarily free to hire experts in crystal growth, for example, even if they worked for a competitor, so long as secret information is not divulged.

Two exceptions to this principle must be considered. The first is the possibility that True Blue's employee has entered into a noncompetition agreement with his previous employer. As overt restraints on workers' ability to pursue their livelihoods wherever they choose, such agreements usually face a hostile reception in court – particularly where the only real concern is preservation of trade secrets, which can be protected without punishing the departing employee in advance. But often they can be enforced at least to some degree.

Sometimes courts will allow this form of pre-emptive punishment even in the absence of a noncompetition agreement, if they view disclosure of trade secrets as "inevitable." This second exception to the usual principles occurs only in narrow circumstances and many courts reject it altogether. But where, for example, a court senses opportunism or bad faith, it may view the former employee as too radioactive to release into a competitive environment – particularly if that environment is limited to one or a few direct competitors and the employee can make a living elsewhere.

Once again, Sandy might do a little homework, beginning with mild inquiries about True Blue employees' former positions, progressing to more focused questions on the nature of their earlier duties, and perhaps an express request to see prior employment agreements.

Finally, delving into the past also includes checking whether any of True Blue's inventions was made outside the United States; True Blue makes no secret of the fact that Dmitri's expertise was honed in Russia, for example. If so, apart from questions of ownership, it is important to ensure that foreign-filing licenses were obtained before the US patent applications were filed.

Has True Blue given anything away?

To dig still further, Sandy can perform various electronic searches: in the United States, the files of many federal and state courts, the Securities and Exchange Commission, and numerous secretaries of state are available online. A litigation docket search will reveal whether True Blue is (or recently has been) embroiled in a lawsuit. The secretary of state in True Blue's jurisdiction will have records of security interests covering True Blue's assets – which, if they exist, should be in the names of the company's bank and other financial creditors. Unexpected security interests can reveal otherwise hidden creditors, *even the sinister Mr. Big!* And, although True Blue's stock is not publicly traded, any significant agreement between it and a public company may show up in the latter's periodic filings with the SEC.

Sandy and his raging suspicions have decamped to True Blue, which owes him a tour of its facility. Located in a one-story building with a corrugated roof and peeling paint, True Blue's equipment and facilities are belied by its modest exterior. The showpiece is a state-of-the-art clean room surrounding a crystal growth reactor. Dmitri Ovkorsky is explaining its operation as Sandy's mind wanders to the round of golf he has scheduled. Rain is forecast but what do they know? Finally something Dmitri says catches on a mental barb.

"Did you say 'open source'?" Sandy asks.

"Yes," Ovkorsky beams, "this is one reason we have been able to come so far with so little funding. Our neural network framework, our webservers, all based on open-source software – and therefore all free."

Sandy hasn't done many software deals, but others at HSELP have, and Sandy often hears them groaning about "open-source problems." Something to check into.

"How is your due diligence going, by the way?" Ovkorsky asks. "Do you have everything you need?"

"Good, good," Sandy says, grateful for the opening. "Have the patent applications been assigned yet?"

"Of course," says Ovkorsky.

"Oh. Well, we haven't seen the documents yet."

"Wait here, I'll make you copies. Apparently my lawyer forgot to include them in the files he sent. It didn't occur to your lawyer to ask why they were missing."

"He checked the records at the patent office –"

"Which are months behind," says Ovkorsky. "Not very surprising. How long did your town's registry take to record the deed to your house? Mine took a year."

"I guess."

"Perhaps your lawyer wasn't that concerned."

"Perhaps not."

Sandy bides his time wandering around the clean room, a box-like structure of steel trusses and acrylic windows, a humming room within a room. Inside, workers in spiffy white head-to-toe hazard suits fiddle busily with the refrigerator-sized reaction chamber. As he walks around the transparent enclosure, Sandy notices a legend etched into one of the panels. It's been scratched out. Stealthily Sandy eyes the legend as he pretends to watch the activity inside. Dmitri is coming back, copies in hand. The inscription appears to be a long serial number followed by – Sandy is almost sure – the words "US Army."

"Nice reactor," Sandy says with nonchalance.

"On loan," Dmitri replies.

Sandy is surprised. "I've heard of renting power tools and floor polishers, but semiconductor fabrication equipment?"

"We are actually beta testing it for the manufacturer. It's a new model. We fine-tune our recipes, they get performance reports. We're both happy."

"Can we see the beta agreement?"

"Of course," says Ovkorsky.

Dangerous liaisons

It is critical, in any due diligence, to determine whether the target company has "in-licensed" any of its technology from external sources, or "out-licensed" it to outsiders. Such licenses can range from the routine – end-user license agreements (EULAs, as they are often called) for commercial software, for example – to the highly particular. All out-licenses must be considered carefully. Even a seemingly innocuous EULA given to customers can contain potentially extravagant obligations or expose the licensor to unreasonable risks. For example, the Uniform Commercial Code presumes that goods (including, in most places, software) are sold with certain implied warranties, such as a warranty against IP infringement. Unless expressly disclaimed, implied warranties are automatically read into a contract.

In many cases, the seller's exposure for breach of warranty is limited to the purchase price of the goods: all is forgiven if the seller issues a refund. But not always. If the buyer suffers some additional harm, for example, lost profits due to the sudden withdrawal of the goods, the buyer can sue for damages beyond the purchase price. Particularly in transactions involving software, the difficulty of identifying competing IP rights frequently motivates sellers to disclaim the warranty against infringement altogether, or at least responsibility for "consequential" damages above the purchase price.

Sophisticated buyers, on the other hand, especially when negotiating large purchases from new or small sellers, will not let them get away with that. They want the warranty and more, often demanding an indemnification against IP infringement. IP indemnification provisions should strike fear into the heart of any potential investor. Without limitations on the seller's financial exposure, it could face a ruinous obligation to defend customers against lawsuits that arise unforeseeably and evolve unpredictably. British Telecom, for example, filed a lawsuit in 2000 claiming a patent applied for in 1977 covered hyperlinking – which, of course, did not even come into existence as we know it until much later. Broad patent claims can cover quite a bit of still-uncharted territory and technical terminology changes rapidly, making it all but impossible for anyone to locate every IP right that could be asserted.

The general principle, when evaluating licenses, development and joint-venture agreements, or even seemingly routine distribution agreements, is to look for terms that compromise the future. It is a sad fact of life that young, unfunded companies often strike deals – whether out of eagerness or naïveté – that trade exclusivity for the prestige and/or cash of a relationship. Prematurely committing the company's fortunes to a single partner, often on questionable terms, can limit business strategies. Sometimes the best service a venture capitalist can provide to a prospective portfolio company is a strong hand in renegotiating agreements before closing. Although the company may chafe at the prospect of re-opening what was probably a painful dialog, resentment will give way to gratitude when the other side learns that professional investors cannot live with imprisoning terms, and that the alternative to a realistic deal is a bankrupt partner.

Even arrangements having only tangential relevance to IP can involve compromising commitments. True Blue's beta agreement with the reactor manufacturer, for example, may compel True Blue to provide test results – that is, after all, why manufacturers enter into beta agreements – with sufficient specificity to divulge its recipes. Absent appropriate confidentiality safeguards, such information can become public. Of even greater concern is the proclivity among beta providers to claim ownership of IP developed using the beta product. To a point this is understandable: no manufacturer can permit development partners to own discoveries they make about its product, lest those partners turn the tables on the manufacturer and demand tribute for their use. This legitimate concern, however, does not justify overreaching – for example, seeking a foothold in the user's own IP by means of an overly broad reservation of IP rights. The manufacturer, in other words, may be entitled to any discoveries True Blue makes about the quirks and limitations of its reactor, but cannot be allowed to own, say, True Blue's recipes merely because True Blue devised them using the reactor – especially if nothing about those recipes is tied to use of that *particular* reactor. Sandy must peruse the beta agreement to ensure that True Blue has not permitted leakage of its trade secrets or signed away any IP rights inadvertently.

The use of equipment or materials as bait to troll for IP rights is hardly limited to beta-testing arrangements. Particularly in the life sciences, manufacturers of patented or difficult-to-make biological products may release such products only pursuant to "material transfer agreements" that claim rights to *anything* the user discovers. Though that seems like a pretty bad deal for the user – why bother to perform research someone else will own? – young companies or uninformed researchers may enter into such grantback agreements without adequate thought, delighted to exchange speculative rights for tangible enticements.

The pleasures and terrors of open source

When professional investors examine a potential portfolio company, the scent of "open-source" software can cause an allergic reaction. Although such software has been around for decades and represents perhaps the fastest-growing category of software worldwide, the obligations imposed by open-source licenses have only recently attracted widespread attention – and most of that negative, at least in the investment world. Adopt even a thimbleful of open-source code, so goes the fear, and invite a flood of disclosure obligations that will wash away proprietary rights in your entire system.

While overstated, the fear is far from irrational. Open-source obligations, when they exist, limit the user's ability to enforce private rights. But the allure of open source is considerable – readily available, tested software to perform routine or specialized chores, often at no cost. Telling developers to spend months reinventing what they can download in a moment offends them as absurdly wasteful. Most open-source software is well-documented and constantly refined by a worldwide community of committed, bug-sniffing volunteers. Some programs have become *de facto* standards. These temptations can blind otherwise careful engineers to the risks; how can something so pleasant and easily ingested, they may wonder, make anyone ill?

Complicating matters is the sheer number of licenses out there. Open-source software is not always free of charge and is almost never free of obligations and restrictions. An important basic principle is "copyleft": in contrast to traditional copyright, which gives the owner of a work the exclusive rights of use, modification and distribution, open-source licenses give the user free rein. Copyleft means that you must pass on to others the same freedoms that were available to you.

At the same time, copyright is hardly foreign to the open-source scheme. In fact, copyright law and its notion of a "derivative work" form the axis on which the open-source world turns. A derivative work arises when an original work is changed in some way that does not alter its essential character (for example, translated from one language into another) or when it is absorbed into something new. As long as a substantial part of the original work persists in the new work, it does not matter if the original forms only a tiny part of the new; a short open-source subroutine slipped into a massive system program in effect produces two derivative works: one of the subroutine, and one of the subroutine-free system program. Copyright law respects both perspectives. Pin a moustache on the Mona Lisa and you have, on one hand, a Dada desecration of Leonardo's masterpiece, and, on the other hand, that moustache set against a colored background.

Absent permission in the form of a license, only the owner of an original work may make derivative works. Which means that, if you incorporate open-source software into your product but fail to comply with the relevant terms of use, you are a copyright infringer. Copyright makes it almost impossible to avoid licensing obligations when open-source software is used in any way.

Although open-source licenses are legion, the most widespread (and widely misunderstood) agreement is the General Public License (GPL), which covers all sorts of open-source programs. The GPL is also among the most onerous of open-source licenses; others typically limit proprietary rights to a lesser (and often far lesser) extent.[1] A program that incorporates code subject to the GPL is, as a derivative work, itself covered by the GPL. Thus, GPL-covered software is "viral" in the sense that, even if minuscule, an open-source component will "infect" (that is, make subject to the GPL) the entirety of any program that contains it. Hence investors' fitful reaction.[2] What obligations does the GPL impose? First, it prohibits royalties or other license fees based on use. But nothing prevents a developer from charging a one-time purchase price (in effect, a delivery fee) for programs subject to the GPL. The developer may also charge for warranties, support, service, updates and revisions, and indemnifications. The GPL covers none of these extras.

Second, the developer must give purchasers the software's source code or the right, exercisable for 3 years, to receive it for no more than the cost of its distribution. This is where any possibility of maintaining proprietary rights becomes untenable. In addition to furnishing source code, the developer must allow customers to freely modify the program, and then (pursuant to the "copyleft" principle) to distribute it to others on the same GPL-dictated terms.

[1] For example, the "lesser" GPL (LGPL), another open-source license, usually permits users to maintain proprietary rights in their own program code. The LGPL distinguishes between dynamically linked libraries, which are called only when needed and remain separate from the main program, and statically linked libraries that are swallowed whole and form part of the program's execution code. If a program links dynamically to a library covered by the LGPL, a derivative work has been created, and the LGPL applies. But the obligations are minimal.

[2] On the other hand, works merely "aggregated" with material subject to the GPL – for example, supplied on the same CD-ROM but not incorporating or becoming integrated with the open-source code – lie outside the GPL's reach.

The prospect of sacrificing proprietary rights may seem ludicrous, but, as a business strategy, often it can make sense. The very transparency of open-source products may provide customers with far greater comfort than a walled-off system, particularly one sold by a small, potentially vulnerable company. The prospect of a worldwide community of users correcting errors and patching security flaws, free of charge, can also hold great appeal. So long as customers (and others likely to encounter the source code) are unlikely to become competitors, a deliberate open-source strategy may be quite viable.

What if a user disregards open-source obligations? Will he be smitten by a mighty source of wrath? So far, despite its elaborate terms and reasonably long history, the GPL has never been tested in court. Moreover, there is no central enforcement authority; while the Free Software Foundation (FSF) wrote the license, it is generally not a party to lawsuits involving open-source software unless its own rights are at stake – for example, if FSF itself created the software in question. FSF characterizes the "free" in free software as referring to freedom, not costless use. ("Think free speech," they explain, "not free beer.") To this one might add, at least in the case of the GPL, that there is no free lunch for those averse to becoming part of the meal. But the extent of open-source obligations is not monolithic, and only becomes an issue when used in conjunction with products sold commercially.

So rest easy, Sandy. True Blue uses its software internally; it never becomes part of a product they sell. Open-source obligations, no matter how oppressive, should not affect its proprietary rights.

Five-putting on the sixteenth green – it hadn't rained after all – Sandy tries to focus on True Blue's beta agreement with the maker of the reactor, but he can't tear his thoughts away from that immaculate, airy clean room. His exertions have turned up nothing – no litigation, no security interests, no apparent source of funding for that fancy equipment. In trying just to estimate the enormous cost, Sandy is becoming an expert in air filters and humidity controllers. It came from the Army, that much is clear. The only question is how.

The whole setup is too neat. All of True Blue's inventors have signed employment agreements and none has a questionable past relationship. Open-source problems have been neatly sidestepped. The company's entire history could have taken place inside that clean room, it's so spotless. Which can mean only one thing: this Ovkorsky is one diabolically clever character. Probably he's feeding the CIA everything he knows about Russian military research and his payoff took the coin of fancy equipment.

No, Sandy realizes, taking his eyes off the little white ball. Too simple. If the clean room came legitimately from the US government, why the amateurish attempt to obscure its origin?

The ball sails past the hole as the picture becomes all too clear: Dmitri must have gotten it from the Russians in some fiendishly complicated deal – desperate to jumpstart his research, he gave the Russians something so valuable they took his shopping list and lifted a high-end clean room with HEPA filters and ceiling grids, humidity control and tear drop lighting *out from under the noses of the US Army!* An astonishing feat. Sandy can't decide whether to lead the investment syndicate or call the FBI.

Government funding

Let us suppose, just suppose, that True Blue came into its high-end government equipment through official channels. If the government has a hand in funding True Blue, whether by way of cash or capital equipment, what rights does it obtain? Can it limit True Blue's ability to pursue its business and enter into exclusive relationships?

In 1980 the United States adopted an explicit policy[3] of allowing universities and businesses operating with federal contracts to retain full ownership of government-funded inventions – including the right to obtain and own patents – for the purpose of further development and commercialization. Contracting universities can exclusively license the inventions to manufacturers; contracting businesses can do the same or commercialize the technology themselves. The rationale is to stimulate the domestic economy through ultimate manufacture of products in the United States. All the unversity or business must do is formally elect title to the invention no more than 2 years after its disclosure to the government.

The federal government does not step out of the picture entirely, however; it retains "march-in" rights, which kick in if the invention is not being made sufficiently available to the public. March-in rights permit the government to license the invention to another company, without the consent of the patent holder or the original licensee, in order to get things moving – to issue, in other words, a compulsory license. The government also gets a royalty-free, non-exclusive license to use the invention for official purposes (including use by government contractors). These modest, mostly theoretical rights rarely raise due-diligence eyebrows; if they come into play at all, it is because the electing company (or licensee) – the entity of concern to a prospective investor – has lost interest anyway. While it is important to review the funding contract to verify compliance with disclosure and election requirements (since failure to comply can strip the university or business of title permanently), government

[3] Embodied in a federal law called the Bayh–Dole Act.

rights, at least in the United States, generally should not worry prospective investors.

In other countries it is a different story. Few have adopted as aggressive a policy to privatize government-funded inventions; instead, most view such inventions as a public trust, and it is therefore critical to investigate both the nature of the government contract and its consistency with national law, which may well override it. Moreover, many countries exhibit far less hesitation in granting compulsory licenses to a technology developer's local competitors if, for example, the developer fails to get a product to market fast enough, or at a low enough price (in the view of the government) – *whether or not the government provided funding.* The threat of compulsory licensing, in such countries, applies to every issued patent. World Trade Organization rules place some limitations on compulsory licensing practices, but it is important to investigate them (particularly in the case of pharmaceuticals) for important foreign markets.

"Let me ask you something, Dmitri," Sandy says, trying not to betray the edge in his voice. "Who are you – that is to say, True Blue – most afraid of?"

"Competitively? Well, as you know, there are some big players –"

"Not competitively, necessarily," Sandy says carefully. There's a pause and an odd noise in the phone, as if Dmitri is fidgeting with his handset.

"Ah," he says at last, "you mean whether we can be threatened. Whether someone has the goods on us, yes?"

"Exactly!" Sandy exclaims, and it all comes spilling out. "Come clean, so to speak, Dmitri! I can hear the hum of those HEPA filters! Where'd you get 14 million dollars' worth, give or take, of high-end isolation equipment?"

"Oh, that," Dmitri replies with apparent relief. "This is something I merely can't tell you. I was afraid you were concerned about competitive patents."

Sandy feels himself reddening. "What competitive patents?"

"We swim with some very big sharks. Sharks covered with more patents than scales."

"Yes, yes, of course. I knew that."

"Well, I'm pleased you aren't worried. Most VCs probably would be. But here all you're concerned with is the origin of some equipment."

"Which is?"

Ovkorsky is positively mirthful. "If I told you that," he chuckles, "I'd have to kill you."

Sandy really should have thought about competitive patents much earlier. An expedition to locate them takes time and money, and this expensive journey – the ultimate outward-looking due-diligence task – should not be undertaken lightly. If a patentability search is like a biopsy, a "freedom-to-operate" or clearance search is more like open-ended exploratory surgery. Whereas a patentability search canvasses the literature for subject-matter relevance and largely ignores the claims of prior patents, a freedom-to-operate search focuses almost entirely on patent claims. Moreover, while duplicative or cumulative references can be disregarded in the course of a patentability search, the clearance searcher must not only gather every patent currently in force that may have possible relevance, but must venture beyond the "usual suspects" to scrutinize patents that may seem irrelevant but have broad claims. She should also provide copies of published but as-yet-unissued patent applications. In general, busy searchers expect a two- to three-week lead time before delivering raw search results. They usually work within an initial budget ($1,500–2,000 is common) and report back if additional search time is likely to yield further results. A patent attorney reviews the patents and published applications located in the search, and supplies an overview of the competitive picture. The total cost is rarely much below $10,000 and can easily climb.

An initial evaluation of a project should determine whether outward-looking due diligence will be important and, if so, prompt early planning for a clearance search. Part of this planning involves recognition of any search's limitations, and part centers around cost management. The grim reality is that clearance searching involves a cascade of guesses, and the error probabilities of all the guesses combine to produce, in the end, a more uncertain guess. Still, the exercise is not without usefulness; just do not expect more risk reduction than the process can feasibly deliver.

The first task is to identify the product or process features of greatest interest. Like patentability searches, a clearance search can cover only one or a small group of highly related features. Unfortunately for the due-diligence consumer, the typical product contains several features that might impart a market advantage, and what is worse, even *uninteresting* features can infringe competitors' patents. The first guess, therefore, derives from the inability to search every potentially infringing feature. Pick the important ones and live with that.

The second source of guesswork lies in evolving technical vocabularies. Consider once again British Telecom's claim that their 1970s-vintage patent covered hyperlinking. Too bad the term did not appear in the patent or even enter common parlance until well over a decade later. If a cautious Internet service provider wanted to test the IP waters before plunging in, it is inconceivable that its search efforts would have turned up British Telecom's patent.[4] Indeed, even current technical terminology can vary among companies and

[4] On the other hand – and fortunately for Prodigy, the first defendant – the court disagreed with British Telecom's contention. So perhaps the likely invisibility of this patent to search efforts was correct after all.

regions. So no clearance search, regardless of the budget, is likely to turn up every issued patent.

Then there are the *un*issued patents. Patent applications usually reach the publication stage long before grant. The claims in such applications are merely entreaties for protection and may have been written in ignorance of the prior art. But there they are, in print, covering tremendous swaths of territory. The psychological burden shifts: having discovered what they perceive as a palpable threat, investors may demand proof of harmlessness as if the patent had already issued. That is an overreaction, but concern is surely warranted; someone has, after all, staked out a claim to critical strategic rights. Soothing words will not coax the genie back into the bottle. But an overview of the prior art (which may well be supplied by the clearance search itself, perhaps supplemented by a patentability search that includes expired patents), demonstrating that broad rights will likely be unavailable, can at least curtail his potency.

And then, finally, there are the patent applications that have been filed but as yet remain unpublished. These applications represent the "dark matter" of the patent universe – invisible, unsearchable, and of unknown extent. The eighteen-month lag between filing and publication, therefore, represents still another source of uncertainty. A competitive application could publish the day after the search is performed and thereby elude discovery. Clearance searches are not easily updated; although the patent office classifies each application according to a massive list of technology categories, the typical clearance search covers many such categories. It may be productive, if substantial time passes between the search and closing, to check the most relevant ones for intervening publications. But, depending on the interval and the categories involved, reviewing them can itself represent a substantial undertaking.

Given the cost and time involved in a clearance search, it may be tempting to cut corners in some fashion – for example, by confining the search to patents owned by known competitors or issued within a specific time frame. Such artificial limitations can dramatically reduce the value of an already inherently imprecise process. A search limited to competitors, for example, ignores many published applications (which have not yet been assigned or had their assignments recorded), patents attributable to key inventors before they joined a competitor, patents still registered to a predecessor entity (which may have changed its name, merged, etc.) or to a holding company for tax purposes, patents licensed to but not actually owned by a competitor, university efforts, and work by large companies that only dabble in the area but whose dabbling produces reams of patents. If clearance is called for, do it right or do not bother; a false sense of security is far worse than a business risk undertaken with open eyes.

When, then, is clearance called for? The less time a product has been on the market and the greater the quantity of new features, the more a clearance search should be considered. Some also consider the size differential between the producer and its competitors, but, in truth, even in the hands of an individual a patent can – with the help of litigators who work on a contingency basis – prove as deadly as one owned by General Electric. True Blue has no products in the marketplace, lots of features it considers novel, and enormous competitors; evaluating its freedom to operate is clearly wise. Indeed, many companies routinely perform clearance searches before seriously exploring a new invention, recognizing the relatively small cost of the search compared with that of product development.

"Tell him where you got the clean room, Dmitri," True Blue's lawyer deadpans, "before he calls the feds."

"I wasn't going to –"

"I know, I know. Cool it, Sandy," the lawyer says. "Dmitri is afraid of having this information leak out. But I said he could trust you. I'm right about that, Sandy, am I not?"

"Certainly."

"I assured him we have a strong nondisclosure agreement in place, and that I'd hand him your butt if word got out. Right, Sandy? So go ahead, Dmitri."

Dmitri's studied glare has something impish behind it – the look of a prankster annoyed at being caught but pleased with the cleverness of it all.

"We purchased it," he finally admits with a sigh, "on eBay."

Sandy eyes him skeptically. "eBay?"

"Military surplus."

"And you paid . . ."

"Fourteen dollars and ninety-five cents. Including the hazard suits. Some real bargains out there."

Sandy's jaw drops. "Fourteen *dollars?* For all this?"

"I was the only bidder. The listing just said 'clean room.' Maybe people thought that sounded high for a little sprucing up."

"But fourteen *dollars?* Not fourteen million?"

"That's why we don't want the word to get out," the lawyer says. "There are other items on our shopping list, and, if the media gets hold of this, it may embarrass some bureaucrat into more, shall we say, unfavorable pricing."

"Anyway, it was almost fifteen dollars. And besides," Dmitri says, still annoyed, "the shipping was murder."

IP due diligence checklist

No discussion of IP due diligence would be complete without the obligatory checklist. So here you go. But do not approach a project by resorting to the list and ticking off the fetching items. That would betray Nietzsche and his injunction to plan. Instead, consider the company broadly and narrow your focus gradually. What role will IP play in executing the business plan? Which is more important to the company – a strong offensive IP enforcement position or freedom to operate? (Do not say "both." It's never both in equal measure. The weighting depends on the nature of the company's innovation, its relationship to the market, and the size and strength of its competitors.)

Build the outline first. Fill in as much as you can with specific tasks. Then use the list as your palette to complete the fill, making sure nothing is left out.

Be prepared to deviate. Although a due diligence usually does not turn into the paranoid odyssey that Sandy experienced, the process can take unexpected turns that demand exploration. It's important to approach problems constructively, with a view toward solution, preferably before a transaction takes place or is even agreed upon. Unpaid fees can be submitted, assignments obtained and recorded, omitted references cited, and troublesome open-source components replaced with home-grown alternatives.

First things first

- Obtain and review business plan.
- Identify key technology drivers, differentiators from competition.
- Identify primary competitors.
- Determine if products on sale. Have they been publicly disclosed or used? Ascertain dates.
- Evaluate company's IP strategy: patents vs. copyright vs. trade secrets; foreign protection program.
- Evaluate company's IP procedures for identifying, evaluating and protecting inventions.
- Perform SEC search for IP-related company filings or third-party filings mentioning company.

IP Assets

Patents

- Obtain a status list of all patents and patent applications worldwide.
- Verify accuracy with electronic search.
- Determine if foreign rights been preserved.

- Review patents, patent applications, analyze claims; for important cases, review file histories.
  - Do claims cover company's products?
  - Do claims advance the business plan?
  - Any weaknesses? Can they be designed around?
  - Can likely claim coverage be predicted? Is prosecution strategy sound?
- Ascertain whether any applications are involved in oppositions or re-examinations.
- Verify that all maintenance fees, annuities have been paid.
- Assess cross-citation of prior art among related or similar applications.
- Ensure small entity fees appropriate if selected.[5]
- Patentability studies
  - Review patentability opinions obtained by the company.
  - Consider performing patentability search on key features.

Copyrights

- Identify copyrightable subject matter, particularly software, and determine if registered.[6]
- Obtain list of copyright registrations, and review them.
- Verify accuracy of list with electronic search of Copyright Office records.
- Determine whether proper copyright notices and legends accompany literature, instructional material, software distributed to customers.

Trade secrets

- Determine whether company relies on trade secrets, and, if so:
  - Determine what they are, and whether they are more appropriately protected by patents.
  - Determine what procedures company employs to guard against theft (site security, access restrictions, document and computer security).
  - Review employee exit procedures.
- Review employment and consulting agreements for proper confidentiality provisions.

[5] The US PTO gives individual inventors and companies qualifying as "small entities" a discount on official fees. But it is up to the applicant to ensure, each time a fee is paid, that small-entity status remains appropriate; improperly paying the reduced fee can, in some circumstances, result in invalidation of the resulting patent.

[6] Registration can be important in the United States; seldom is it necessary, or even possible, elsewhere. See chapter 1.

- Review noncompetition agreements for enforceability.
- Review nondisclosure agreements company has used with third parties for important collaborations or relationships, ensure they were actually signed.
- Investigate whether company has been sued in connection with others' trade secrets, or has sued others.

Trademarks

- Obtain a status list of all trademark registrations and applications for registration worldwide.
- Obtain and review trademark searches performed by company, attorney opinions concerning them.
- Ensure that all registered trademarks have been renewed, and that all necessary filings have been made to keep the registration in force.
- Determine if ® symbol is being used, and, if so, whether that mark is in fact registered.
- Determine if any oppositions or cancellation proceedings have been filed against applications for registration.
- Verify proper identification of trademarks in literature, on packaging.
- Investigate policing efforts to prevent marks from become generic.
- Investigate any domain-name disputes.

IP ownership

- Perform title search to confirm ownership of all patents and patent applications, trademark registrations and applications for registration, registered copyrights.
- Investigate past employment, recent academic affiliations of key inventors.
- Learn whether core members of development team are still employed at company.
- For copyrights, determine:

  For employees, whether employment agreement covers coyprights.

  For consultants, whether consulting agreement not only characterizes work product as "work for hire" but also requires express copyright assignment.

- For patents, determine whether employees and consultants have a contractual obligation to notify the company of inventions, to assign them, and to provide any further assistance needed to secure patent rights.

IP encumbrances

Out-licenses

- Will rights interfere with company's future? Dangerous exclusivity, for example, or excessively broad scope?
- Transferability restricted?
- Breached by licensee? Profitable?
- Grantbacks reserved?
- Has licensee declared bankruptcy?
- Trademark licenses: adequate provisions for quality control?

Government rights

- Have title-retaining elections been properly and timely made?
- Have government contracts been fully complied with?

Security interests

- Perform UCC search for security interests in IP.

Other agreements

- Review any other agreements implicating IP rights for potentially harmful provisions (exclusivity, IP grantbacks). Examples:
  - Joint development or collaboration agreements
  - Distribution agreements
  - Material-transfer agreements
  - Confidentiality agreements

IP liabilities and third-party rights

- Identify all IP-related lawsuits brought by or against company, all IP-related threats (oral or written) received from third parties.
  - Perform electronic litigation docket search
  - Review judicial opinions and orders
  - Review settlement agreements

- Identify all offers of IP licenses received from third parties.
- Identify and analyze all in-licenses.
    - Are granted rights (definition of licensed product or process, field of use, territory, duration) broad enough?
    - Transferable without consent of licensor?
    - Onerous grantbacks?
    - Has licensor declared bankruptcy?
- If possible, review all opinions of counsel relating to infringement of third-party IP rights and/or third-party infringement of company-owned IP.
    - If not possible to review, at least learn and assess reasoning.
- Consider freedom-to-operate search for competitive patents.
    - Review any freedom-to-operate searches performed by or for company, and, if not possible to review opinion, at least obtain references considered.
- Assess employment history of key innovators, review their prior employment agreements. Any basis for possible misappropriation claims from third parties?
- Open-source issues
    - Identify open-source components in company's *operations* as well as its *product offerings.*
    - All license obligations complied with?
- Standards issues
    - Does company implement any industry standards?
    - If so, are license agreements in place?
    - If company has contributed to an industry standard, did it comply with standards organization's IP rules?

7 Licensing and related transactions

Licensing is the vehicle by which the IP rubber meets the commercial road, with someone else doing the driving. The idea of allowing others, even competitors, access to hard-won proprietary rights may have seemed absurd a generation ago, but today it is commonplace and a huge source of IP-driven revenue. To be sure, licensing is by no means the only way to realize value from IP; many patents thrive locked away in a safe, securing the realm against marauding technology thieves and increasing (sometimes largely constituting) a firm's asset value. But, as soon as a company or institution finds it attractive to look beyond its own capacity to monetize technology, licensing becomes inevitable.

A license is essentially a promise not to sue. You are given permission to infringe someone's IP without fear of legal repercussion. A ticket to a concert is also a license – permission to enter and listen without being thrown out unless you make too much of a pest of yourself (that is, violate the terms of the license). Unlike a concert ticket, however, an IP license is *personal.* Concert promoters do not care if you give your ticket to someone else. An IP owner, by contrast, does not want valuable rights falling into the wrong hands – those of her competitor, for example, or hands simply less capable of effectively exploiting the IP (for the ultimate benefit of its owner). Most license arrangements reflect a balance between the licensee's need for enough rights and incentives to justify the market risk of product introduction, and the licensor's aversion to tying up or losing control over its innovations.

Basic terminology

Owners have broad latitude to carve up the IP turkey as they see fit – dictating when, where, and to what extent a licensee is free to exploit the rights it receives. An ***exclusive*** license empowers the licensee and no one else; even the licensor cannot make use of what it has granted. If the licensor wishes to reserve the prerogative to make and sell the technology and compete with the licensee in the marketplace, it grants a ***sole*** or ***co-exclusive*** license rather than an exclusive one.[1] Because an exclusive license cedes so much control to the licensee, it may be treated as a complete transfer of ownership for tax and other purposes. An exclusive licensor, fearful of entrusting its proprietary rights to a frog disguised as a prince who will botch their exploitation or forget about them entirely, typically insists on various performance measures and rights of termination.

A ***nonexclusive*** license, by contrast, can be granted to numerous takers on the same or varying terms. A nonexclusive license may seem easy to grant, since it does not preclude other such licenses. But it does eliminate the possibility of granting *exclusive* rights to anyone. A nonexclusive license commits the licensor to a one-to-many rather than a one-to-one strategy. If the licensor has second thoughts (or believes, incorrectly, that the nonexclusive license has expired) and tries to grant an exclusive license notwithstanding the earlier commitment, the licensee is out of luck – even if it did not know about the prior license – because a licensor can grant no more than it owns. Licenses are typically confidential, so any prospective licensee must assure itself it is receiving what it bargained for.

A license may be ***royalty-bearing*** or ***paid-up***. Software publishers, for example, grant paid-up non-exclusive licenses to their customers, because in purchasing the software they have, in fact, paid up. But money need not change hands for a license to be paid-up. A ***cross-license*** allows two patent owners each to make use of the other's patents, often without any form of balancing payment. Litigation settlements, for example, often include cross-licenses. Paid-up licenses are often ***irrevocable*** – no matter what the licensee does, the license is hers forever – and ***perpetual***.

A ***grantback*** may also be couched in the form of a paid-up license, either exclusive or nonexclusive. A grantback provides the licensor with rights in any improvements or modifications that the licensee makes to the licensed technology. That may seem rapacious on the part of the licensor, but, in fact, grantbacks enjoy a long tradition and reflect the licensor's legitimate aversion to being walled off from its own technology. If a licensee figures out a way to make the licensed subject matter faster, better, or cheaper, it has done so only because the licensor has given it access in the first place; without a grantback, many licenses would never be signed, so great is the fear of being leapfrogged.

The terms of a grantback may vary in terms of exclusivity both with respect to the licensor and the licensee: the licensor may seek an exclusive grantback, preventing the licensee from allowing others to use its improvements, while

[1] This is United States terminology. In other countries, an "exclusive" license may correspond to what Americans call a sole license.

the licensee may seek restrictions that prevent the *licensor* from re-licensing the grantback to others. United States law tolerates nonexclusive and (to a lesser extent) exclusive grantbacks, as well as outright assignments of improvements, so long as the effect is not anticompetitive. As we will soon see, other countries draw the line at the nonexclusive variety.

Rights may be worldwide or limited to a defined territory, of indefinite ("perpetual") duration or restricted to a fixed term, unlimited in scope or confined to a specified field. As explained in chapter 3, field limitations allow the licensor to have different companies pursue different applications of its technology, and those applications may be segmented – by industry (or, more finely, within an industry), by market, by customer identity or type, according to the type of equipment to be sold or the use to which it is put, therapy v. diagnostics, medical v. veterinary – in any way that makes business sense, so long as the licensor can find different licensees willing to coexist.

Because licenses are personal, they cannot be ***sublicensed***, allowing the licensee to delegate the granted rights to someone else (either in lieu of or in addition to itself), without the consent of the licensor.

Basic terms

The most important terms of a license specify the nature and duration of the licensed rights, how the licensor will be paid, and the conditions under which either side can extricate itself from a deal gone sour. Exclusive licenses focus heavily on the latter issues, and also specify the efforts the licensee must make to bring the technology to market. It is often wise to obtain a common understanding of these basic underpinnings by way of a term sheet before drafting the full agreement.

The layers of complexity built on this foundation reflect planning for contingencies and the unexpected, the need to preclude funny business (such as arbitrage), and an exchange of assurances that each side means what it says.

CASE STUDY #5: Let's return to our friends at RSS, who, you may recall from chapter 3, have decided to license their sequencing technique to a large manufacturer of biotechnology. They have struck a deal with GeneMachine, Inc. What basic provisions will they need? What concerns motivate each side? And how might they approach negotiation?

The grant defines the character, scope, duration, and territorial extent of the license. The typical grant allows the licensee to commercially exploit, in any way it chooses, all technology falling within the license. But there is no magic in this formulation; the licensor is free to divide those rights among different recipients, or reserve some of them for itself. For example, a manufacturing license may provide only the right to make, while a distribution license allows no more than sales of already made goods. A garden-variety nonexclusive license grant might read:[2]

RSS hereby grants to GeneMachine for the TERM a nonexclusive license, without the right to sublicense, under the PATENT RIGHTS to make, have made, use, sell, offer to sell, lease, and import LICENSED PRODUCTS in the FIELD in the TERRITORY and to perform LICENSED PROCESSES in the FIELD in the TERRITORY.

Let us unpack that jargon. The term of the agreement may run anywhere from a fixed number of years to the point at which the last of the relevant patents expires, but, of course, the end may come early if either side wishes – and is permitted under the agreement – to pull out. The incantatory language "make, have made, use, sell, offer to sell, lease, and import" is intended to cover all the ways a licensee might commercially exploit patent rights; if any of these is omitted from the grant, it could limit GeneMachine's freedom of action.

GeneMachine, however, will probably insist on an *exclusive* grant providing *worldwide* rights; otherwise, the project is unlikely to be attractive. The typical nonexclusive license is a toll: anyone looking to enter the business must deposit the right change in order to proceed ahead, and can expect to compete with all the other toll-payers. IBM, for example, runs a very lucrative tollbooth. Entering into an exclusive license, by contrast, is more like renting the entire turnpike. GeneMachine wants to keep everyone else out in order to justify the expense and risk of developing, then marketing, a new product – particularly a complex one that will doubtless involve *a lot* of development and intricate production requirements. Without worldwide rights, which RSS has hopefully preserved by filing PCT or foreign patent applications, the size of the accessible market may be insufficient to support those efforts. A suitable exclusive grant might look like this:

Subject to the terms and conditions of this Agreement, RSS hereby grants to GeneMachine an exclusive, non-transferable (except as otherwise expressly provided in this Agreement), worldwide license, with the right to sublicense as provided herein, under the PATENT RIGHTS, (a) to make, have made, use, sell, offer for sale, lease and import LICENSED PRODUCTS in the FIELD and (b) to perform LICENSED PROCESSES during the TERM, unless this Agreement shall be earlier terminated in accordance with the provisions hereof.

If the license is to be exclusive, RSS will insist on non-transferability. Most licenses contain at least some restrictions on transfer; the exclusive license

[2] Capitalized terms would be defined with precision in the license agreement.

generally prohibits it entirely. After satisfying itself that GeneMachine can be trusted with its IP treasure, RSS cannot allow GeneMachine to dictate its successor; indeed, many exclusive licenses contain "change-of-control" provisions that cancel the license in the event of merger, asset sale, or other corporate event that fundamentally alters the business identity of the licensee. GeneMachine, on the other hand, will likely insist on some right to sublicense. Expected to exploit the licensed IP to the best of its abilities, and perhaps paying a high up-front fee as well, an exclusive licensee like GeneMachine will resist any constraints on how it does business. That should be acceptable to RSS. Although sublicenses bring others into the picture, they remain subordinate players, and ultimately the obligation to successfully commercialize the technology remains with GeneMachine. What RSS must guard against instead is arbitrage, which brings us to the subject of . . .

Royalties. Fixing a mutually acceptable royalty rate can be difficult, particularly if the underlying economics are to be plumbed; more on that later. For now, it is useful to explore generally how royalties are defined. Usually they are based on "net" sales, meaning that the royalty rate applies to the revenue actually received by the licensor minus a few relatively standard items (for example, taxes, shipping, returns, and certain discounts).

For products involving substantial manufacturing and tool-up costs, royalty rates generally fall in the 1–5% range. But, as with all things contractual, creativity rules, and departures from the simple fixed rate are common. The royalty rate may differ, for example, among products, fields, or territories to reflect varied costs and risks. The rate may be tiered, with smaller percentages applying to progressively higher sales levels, thereby providing greater marketing incentive to GeneMachine. Royalty "stacking" comes into play if GeneMachine must pay royalties not only to RSS, but also to other patent owners on sales of the same product. License agreements typically include a mechanism to reduce the royalty rate in stacking situations, reflecting the fact that the licensor has not delivered the complete set of necessary rights.

Royalty rates for sublicense revenue can be tricky. On one end of the scale is the cheap arbitrage play: the licensee sublicenses to an affiliate at, say, a 1% royalty rate, and the affiliate makes all the sales; if the royalty rate of the main license is 5%, the licensor pockets only 0.05% of sales, while the licensee's corporate cabal retains the rest. That little artifice is easily defeated by simply ignoring the sublicense relationship altogether – basing royalties on the first arm's length sale to an unaffiliated entity, whether or not sublicensing is involved. But what about legitimate sublicensing transactions? Suppose, for example, that GeneMachine sublicenses European manufacture and sales to Quaffing Automation, Ltd. of Tippleshire, England.[3] With the licensee essentially acting as a broker, it is unfair to apply the license royalty rate – which presumes manufacturing and marketing efforts – to income that GeneMachine derives passively from Quaffing's efforts, without expending any of its own. Still, the sublicense relationship represents more than mere arbitrage. GeneMachine did discover and charm Quaffing and thereby open up the European market, and may also provide ongoing support. Some degree of sharing (RSS and GeneMachine may, for example, simply split sublicense income evenly) is called for.

Licenses can include other fees that supplement or reduce royalties. Many licensors expect an up-front payment for the privilege of entering into the transaction; often this payment reimburses the licensor for patent expenses. Particularly in university "spin-out" situations, where energetic entrepreneurs (usually graduating students) attempt to commercialize university-owned patents, the license may also provide for "milestone" payments upon completion of important events – closing of significant financing, regulatory approval, and first commercial sale, for example. University licensors may also obtain an equity interest in the licensee in order to participate more directly in the technology's upside potential and increase the long-term alignment of interest; in return, it may agree to reduce up-front and/or royalty costs. Such arrangements are less likely between a guppy-like RSS and the whale it has hooked, but it happens.

Licensee efforts. In exchange for its willingness to hand GeneMachine the keys to its sequencing technology, RSS will expect guarantees in terms of both effort and result. First, to avoid seeing its IP rights stuffed and mounted for decoration rather than exploited energetically in the marketplace, RSS will demand that GeneMachine use vigorous efforts[4] to develop and introduce commercial products. As an exclusive licensor, RSS will seek minimum royalties. These have nothing to do with the royalty *rate*, but instead set a floor on the yearly income RSS will accept. Minimum royalties do not kick in until enough time has passed for GeneMachine to create and introduce a product, and they typically escalate over time (as sales would be expected to escalate).

A kinder, gentler alternative to minimum royalties is sales targets. Rather than establishing a minimum license income, the targets specify a success

[3] We will not get into it, but note that international licensing arrangements involving the sharing of technical data can require export licenses.

[4] RSS will ask for "best" efforts and GeneMachine will refuse, since that formulation can require extraordinary efforts. The parties will probably compromise on "commercially reasonable" efforts.

criterion. As such, if they are missed, GeneMachine will probably forfeit exclusivity but not the license itself.

Representations and warranties. These are statements of factual assurance that each side provides to the other. They must be true as of the date the license is signed, but, after that, all bets are off unless the license provides for their survival into the future. In addition, some warranties are *implied* unless expressly disclaimed. Implied warranties are well-characterized for sales of goods, but far less so for licenses. Fearing the worst, most licensors include express disclaimers.

RSS's opening position might reject any reps or warranties on either side's part, but it would have little reason to resist a simple set of symmetric warranties confirming that each side has due corporate authority to enter into the license. GeneMachine may insist on more – for example, an assurance that practice of the licensed technology will not infringe anyone else's IP rights. That is when the temperature starts to rise. Such an assurance would give GeneMachine the right to back out of the license, probably obtain a refund of whatever it paid RSS, and possibly charge RSS with responsibility for GeneMachine's entire financial exposure should the representation prove wrong. RSS will plead ignorance of others' rights. GeneMachine will respond that RSS is in the best position to investigate and understand those rights. And so on. Such soft-shoe dance steps notwithstanding, in the end relative bargaining power will determine who represents what to whom.

Indemnities. Actually, it can get worse for RSS than a non-infringement representation. GeneMachine may require an indemnity; that is, an undertaking on the part of RSS to defend GeneMachine against IP lawsuits brought against it. That prospect will rightly horrify RSS, which will resist with all its might, and if it capitulates will at least attempt to limit the obligation – in terms of time (the indemnity obligation can expire after a certain amount of time has passed), financial exposure (for example, capping the indemnity obligation to amounts actually received from GeneMachine under the license), and control.[5] For example, RSS may demand prompt notice of claims and the ability to direct any resulting litigation, which will at least allow it oversight of costs and strategy.

Although indemnities tend to inspire reflexive panic, they can sometimes operate to the indemnitor's advantage by limiting exposure. A blanket

[5] RSS may also disclaim responsibility if GeneMachine could have implemented the technology in a manner that does not infringe the other guy's patent, but failed to do so.

non-infringement representation may, in some circumstances, saddle RSS with whatever pecuniary harm GeneMachine suffers for IP infringement – and that is often a big figure. An indemnity, by contrast, can define a ceiling on RSS's monetary obligation.

Other litigation. What happens if someone infringes the patents RSS licenses to GeneMachine? Responsibility for pursuing pirates generally falls to the exclusive licensee, since it has the most to lose from marketplace competition, but the IP still belongs to RSS; if GeneMachine fails to stop the infringement or bring a lawsuit, RSS will reserve the right to do so. RSS and GeneMachine must consider the mechanics of mutual notification, consent to proposed settlement terms, and decide how the spoils of war will be divided in the event of victory.

. . . and all the rest. Licenses contain plenty of other provisions – termination, reports and recordkeeping requirements, audits, responsibility for patent prosecution and maintenance, and confidentiality, to name a few. These are largely technical and straightforward in nature, rarely stimulating much controversy.

Pegging a royalty rate

An understanding of the economics of licensing is critical to establishing a fair royalty rate. Each side may quibble about the other's emphasis or interpretation, but they should at least share the same underlying assumptions. Unfortunately, because the issue is complex, many grope for the escape hatch of industry averages or rules of thumb. Yet the notion of a "customary" industry royalty is more myth than reality, and, even if one can be identified, it probably will not coincide with what is economically sound under particular circumstances. Adopting an essentially arbitrary figure is likely to favor one side and anger the other in the course of time.

An important component of royalty negotiation, then, is familiarity with the premises behind these shortcut methods and their weaknesses. The most common rule of thumb – known as Goldscheider's rule after the economic damages expert who originated it – holds that, absent unusual circumstances, the royalty rate should provide the licensor with 25% of the licensee's pretax profits from sales. The rule recognizes that royalties ultimately apportion the seller's profits, and that the seller, who must assume the costs of manufacture and the risks of commercialization, usually deserves a larger share than the licensor. For products with pre-tax profits in the range of 15–25%, Goldscheider's rule yields running royalty rates of 3.75–6.25%.

The problem with this shortcut, as well as approaches based on comparable transactions or industry norms, is that general tendencies fail to address the economic realities of specific circumstances – namely, the marketplace advantage conferred by a patent and the resulting "special" profits that may be derived. Consider a patent on a consumer product. The patent allows its owner to exclude perfect replicas of the product as well as some range of alternatives. If acceptable price-competitive, non-infringing substitutes exist, the value of the patent – that is, the special profit attributable to it – is zero. If the patented product is preferred in the marketplace, however, the special profit is the premium the market is willing to pay over substitute products. It is the special profit, not the seller's actual profit, that the parties to a license should fight over. Not that the two quantities can never be equal; if there really is no substitute product out there, every penny of the seller's actual profit is special profit.

Suppose, instead, that the patent covers a production process rather than the item that is eventually sold. In this case, the special profit corresponds to the increased profits retained by a manufacturer using the patented process in lieu of conventional or unpatented alternatives.

Quantifying the special patent profit can be difficult, particularly when no product has yet entered the marketplace. But, for most goods, production costs can be modeled with some sophistication and likely sales prices estimated from existing marketplace behavior. These figures can help gauge the actual and special profits. So now that we have some idea what the parties should be sparring over, how should they split it up? The relevant factors include the degree of risk assumed by the seller, the relevance of the seller's brand in driving sales, capital investments required for manufacture, and any unique expertise the seller brings to bear. The licensee will not adopt the technology unless it earns a fair return, relative to the cost of capital, on the investment and the risks it undertakes. The licensor, for its part, will not license the patent unless it can retain enough of the profit to justify its own investment and to reflect the strength of the patent in excluding substitutes.[6] Patents in the automotive industry tend to have low royalty rates, since individual features do not strongly influence purchase decisions. Customers, in other words,

[6] Unless, of course, the patent is simply sitting on a shelf unused, in which case the licensor may be grateful for any revenue. Funny, though, how the interest of even a single prospective licensee can suddenly inflate the patent owner's perception of value.

probably will not pay much extra for any given feature, so the special profit is marginal. A patent on a pharmaceutical, by contrast, may well offer effective market exclusivity with limited risk and low start-up costs for existing players; the farther along the clinical path the originator has taken the drug prior to licensing, the smaller the risk will be and the larger a share of the profits it can command.

Further complications arise in cases involving patent applications rather than issued patents. The possibility that a patent will never issue, or issue with meaningful coverage, represents another risk factor – but one that can be quantified (at least in broad terms). Prior-art searching and analysis will suggest the coverage likely to be obtained. Market and technical analysis can help place that coverage in context, identifying available and hypothetical non-infringing alternatives.

In the end, because one side's loss is other's gain, some degree of sparring over royalty rates is inevitable – perhaps even healthy. But not if pre-existing prejudices are permitted to obscure the true economic drivers. Licensors, for example, tend to be excessively optimistic about patent strength, while licensees often disdain royalties as a nuisance tax. Ladies and gentlemen, put those prejudices aside, shake hands on a common set of assumptions, and come out negotiating.

The big no-nos

Licensors worldwide have broad latitude to exploit their IP on terms that make business sense. But limits exist. The most important of these involve antitrust liability and IP misuse. Antitrust exposure reflects the desire of trade regulators and courts to maintain a level playing field among competitors. In an earlier day, many courts cast a mistrustful eye on all patents, viewing them as monopolies that inherently tilt the playing field. Today, courts and regulators recognize both the pro-competitive benefits of rewarding innovation through IP protection and the fact that IP itself rarely confers market power. The capacity to exclude competitors becomes meaningful only when the IP covers something the marketplace wants and, without access to the IP, cannot have. When this occurs, owners may be tempted to extend their market power beyond the limited exclusivity their IP rights provide. That's when antitrust laws come into play.

In the United States, two federal agencies – the Department of Justice (DOJ) and the Federal Trade Commission (FTC) – enforce antitrust laws on a national basis, investigating abuses, bringing violators to court, and imposing fines.

Their primary concerns involve agreements or collusion to restrict competition, mergers or acquisitions that have anticompetitive effects, and abuse of dominant market positions. But big federal agencies fry the big fish: the Microsofts and the IBMs and the AT&Ts. Everyone else worries less about the feds and more about competitors, who can file private lawsuits alleging antitrust violations and claiming big-time damages. When sued for violations of IP rights, a defendant can respond with an antitrust counterclaim.

The schemes in Europe and Japan are similar, although the laws and priorities differ. The European Commission (EC) enforces competition law based on the Treaty of Rome, focusing most strongly on preventing trade barriers among the European states. But the EC effectively shares authority with competition agencies in the individual European states, which have their own local agendas. Private firms may institute lawsuits based on local law or, in some countries, to collect damages based on EC enforcement actions (just as businesses in the United States are free to piggyback lawsuits on DOJ or FTC proceedings). Japanese law generally tacks closer to Europe than to the United States, but tends to place greater emphasis on reciprocity of obligations between parties to an agreement; terms that would be deemed unfair if imposed unilaterally may be acceptable if applied to both sides.

Although few companies feel the direct sting of a big enforcement agency, enough uneasiness exists in areas involving IP to prompt the DOJ and FTC in the United States, and the EC in Europe, to issue antitrust enforcement guidelines.[7] These guidelines lack the force of law, but as statements of agency policy, wield considerable clout. Staying within their safe-harbor provisions keeps a company safe from agency enforcement actions, and courts take them seriously when adjudicating private lawsuits.

A concept related to but distinct from antitrust liability is *patent misuse*, which some courts have extended into the copyright sphere as well. Although misuse and antitrust exposure often go hand-in-hand, it is possible to have one without the other. A patent owner lacking sufficient market power to awaken the antitrust laws can still commit misuse, for example. In a private lawsuit, a finding of misuse can deprive the IP owner of the ability to enforce the IP against anyone in the world, while antitrust liability affects just the parties in litigation.

The creative competitor can find many ways to get into antitrust trouble over its IP, but here are the most popular transgressions.

[7] In the United States, these are the 1995 Antitrust Guidelines for the Licensing of Intellectual Property (*see* http://www.usdoj.gov/atr/public/guidelines/ipguide.htm) and in Europe they are the Technology Transfer Block Exemption (*see* http://europa.eu.int/comm/competition /antitrust/legislation/ entente3_en.html#technology).

- *Tie-ins.* Suppose RSS has patents on sequencing machines but, despite all efforts, could not protect the chemical reagents the machines use. Too bad, since the juices are consumables and therefore potentially quite profitable. RSS, undaunted, considers conditioning (or requiring GeneMachine to condition) sales of machines on the purchaser's promise to buy all reagents from RSS. That's a no-no – both on antitrust grounds and as patent misuse – and the more market power RSS has, the bigger the no-no. You can not distort competition by using market power in one area as a lever into another area; that would give you an unfair advantage over competitors in the other area to the detriment of consumers. The law requires you to compete on the merits.

  Now, suppose RSS has patents on both machines and reagents. Can it then condition purchase of machines on a promise to purchase juice from RSS? Maybe. The juice patents give RSS market exclusivity over the covered reagents, so it is not as if the purchase restriction does any more than RSS could do with its patents. But what if non-infringing substitute reagents would work just as well with the machines? In that case, RSS must demonstrate a sound business justification for the restriction (other than the desire to make money). If, for example, there is something special about the patented juice that makes RSS machines work better or more reliably, RSS may be able to justify its restriction, or at least disclaim machine warranty coverage if unauthorized reagents are used.

- *Tie-outs.* IP owners have been known to condition licenses on a licensee's agreement not to compete *in any fashion* with the IP owner. That, once again, represents an unwarranted extension of the IP right – using it to restrict activity not covered by the IP right itself. The practice seems to arise most often in the software context. Particularly if the software is unpatented, its developer, aware that copyright does not extend to ideas, may fear that a customer will learn how the program works, create its own, and become a competitor. Too bad – the developer is stuck with the limits of the protection mechanism he chooses, and trying to escape those limits by imposing additional contractual restrictions may be struck down as copyright misuse.

- *Refusals to license.* In the United States, an IP owner generally has no obligation to deal with anyone with whom it does not want to deal. Even a monopolist may refuse to license a patent. But, if the refusal is really just a tactic to extend market power into an area beyond the scope of the IP, the owner may face antitrust exposure. In addition, a countervailing antitrust principle called the "essential facilities doctrine" has been invoked to require the owner of an "essential" facility (such as a blocking patent or a widely used proprietary platform, such as Microsoft Windows) to share it with competitors. That doctrine has withered over the years and today is rarely available in the United States. Europe and Japan, by contrast, show far less hesitancy in applying it and, in effect, mandating compulsory licensing.

- *Royalties.* There are two excellent ways to get in trouble over royalty arrangements. One is to attempt to extract royalties beyond the expiration of the licensed patents. In the United States that is patent misuse, although probably not an antitrust violation, even though the arrangement might make economic sense – for example, offering a low or zero royalty rate in the early years of a license, when the costs and risks of market introduction are at their peak, and a higher rate extending beyond patent expiration in order to ease the licensee's yearly burden. Doesn't matter – you cannot do it in the US. Europe and Japan are more permissive.

  The other no-no is to use royalties as a means of enforcing a tie-in, e.g., by including within the royalty base sales of products outside the licensed IP rights.[8]

- *Exclusive licenses.* United States antitrust law tends to view an exclusive license as competitively neutral, merely substituting the licensee for the licensor in the marketplace. Antitrust review occurs only in circumstances where a merger of the licensor and licensee would reduce competition, for example, if the parties compete and the effect of the license is to subtract a competitor.

  Europe and Japan also tolerate exclusive licenses but impose greater restrictions – particularly, in Europe, over arrangements that might favor trade in some European countries over others.

- *Geographic and field-of-use restrictions.* While US law interferes little with territorial or field restrictions, so long as the parties to a license are not competitors attempting to divide up the market, European law – concerned, again, with trade parity among European states but not others – imposes more stringent limits.

- *Grantbacks.* Japan frowns strongly on exclusive grantbacks, and in Europe most are flatly prohibited.[9] United States law imposes no blanket restriction, but may prevent arrangements that restrict access to improvements

[8] But basing royalties on sales of unlicensed products is acceptable if those sales provide a convenient and economically valid way of measuring the value of the license.

[9] Europe allows exclusive grantbacks only if an improvement is not "severable" from the licensed technology.

excessively or discourage their creation entirely. Nonexclusive grantbacks rarely raise questions, although European and Japanese law may restrict even these provisions unless the licensor has a reciprocal obligation to share improvements with the licensee.

Licensing and Dealmaking

RSS's licensee, GeneMachine, is in trouble. The market for single-molecule sequencing machines has cooled, and competition is fierce. GeneMachine has entered into buyout discussions with Big Brother, Inc., an international conglomerate with an interest in laboratory equipment manufacture. When GeneMachine asked RSS whether it would object to transferring the license to Big Brother, RSS didn't hide its feelings. First, it has no confidence in their ability to compete in the sequencing market; analytics are only a sideline for Big Brother, whose various businesses revolve primarily around surveillance and eavesdropping equipment. Second, Big Brother appears to be violating RSS's patents on chemical reagents. Seems the juice is useful in systems for detecting airborne bioweapons, and RSS is pretty sure that a Big Brother military subsidiary is selling infringing test kits. So RSS has no inclination to do business with Big Brother; they're liable to wind up in court soon.

But GeneMachine is undeterred; it needs a rescue and sees RSS's refusal to deal as an obstacle. GeneMachine is evaluating its options. What is it likely to find?

Since licenses are personal to the licensee, at first blush GeneMachine has little room for hope. But the situation may be more complicated. The first question is whether the RSS-GeneMachine license prohibits assignment. Let us suppose it is silent on the matter, and that Big Brother has proposed to purchase substantially all of GeneMachine's assets – including the license from RSS. If the license is *nonexclusive*, it will be nonassignable without RSS's permission as a matter of law. Chaos would rule otherwise. For example, Licensees could arbitrage their rights to direct competitors of the licensor, which would quickly lose control of its licensing program and perhaps its business as well. If the license is *exclusive*, however, which is more likely to be the case here, the issue is less clear. An exclusive licensee like GeneMachine may have spent huge sums to build a business around the license, which starts to look more like a property interest than a mere promise not to sue. The law remains unsettled.

Similar principles govern copyright licenses, although here better support may exist for the assignability of exclusive licenses absent contractual restriction. Courts tend to be more permissive about assignment of trademark licenses, which is odd, since the essence of trademark value stems from the nature and quality of the marked goods. You might expect courts to defer to the trademark owner's judgment about the suitability of a manufacturer, particularly when the trademark license contains strict quality-control provisions; but a court may view such provisions less as evidence of the licensor's concern for its reputation than as constraints limiting the judgment and independent skill of the licensee – and therefore the need for the licensor to discriminate. A court may not force a trademark licensor to tolerate assignment to an inferior manufacturer, but the mark owner may have to make the case why its interests would be harmed.

Suppose that, instead of an asset purchase, GeneMachine and Big Brother contemplate a merger with Big Brother as the surviving company. Hear that sound? It's lawyers' hands rubbing with glee at the complications. First, the law is inconsistent as to whether, as a basic corporate matter, a merger should be treated like an assignment. But when an IP license is involved, courts more readily take the assignment view, because the effect on the licensor is the same – suddenly it is dealing with someone new, someone it did not choose. Matters become further complicated when the license contains a nonassignment provision. We know those cover asset purchases, which unquestionably amount to assignments. But such provisions might not cover mergers; a court inclined to preserve a license in such circumstances probably would not be dissuaded by antiassignment language. If the licensor is worried about mergers, therefore, it should use a broader change-of-control restriction that permits the licensor to terminate the license *upon transfer of a majority of the licensee's assets or outstanding voting securities.*

This language covers stock sales, which otherwise would be even less subject to licensor objection than a merger. While an antiassignment provision in a license could conceivably cover a merger, depending on the language and the viewpoint of a particular court, it will not cover a stock sale; only change-of-control language will accomplish that.

Let us say RSS can point to precisely this sort of language in its license to GeneMachine. Big Brother may yet have one more trick up its sleeve. In a normal "forward" merger, GeneMachine merges into Big Brother (or a subsidiary of Big Brother) and ceases to exist. In a stock sale, GeneMachine becomes a wholly owned subsidiary of Big Brother. We know that both of these forms of transaction fall foul of a properly drafted change-of-control provision. But what about that exotic contortion known as a "reverse subsidiary merger"? In a maneuver that could only have been dreamed up by lawyers, GeneMachine does not merge into anyone; instead, a subsidiary of Big Brother – call it Little

Big Brother – merges into GeneMachine, which survives but winds up as a wholly owned subsidiary of Big Brother when the dust clears.

Nice try, but, because the stock of GeneMachine has changed hands, the change-of-control language is triggered just as it would be in a stock sale. The reverse subsidiary merger, like a stock sale, may avoid an anti-assignment provision or the prevailing legal bias against assignment where the license is silent. But it is still a change of control.

Dealmakers, beware. More often than not licenses emerge as insurmountable obstacles to a transaction unless permission for transfer can be obtained. The earlier such obstacles are discovered, the more likely it is that consent will be forthcoming on reasonable terms. Wait until the eve of a much-publicized merger and those terms may become extortionate or disappear entirely.

Licenses in bankruptcy

RSS succeeded in preventing GeneMachine from assigning its license to Big Brother. Now GeneMachine has filed for bankruptcy protection, and the trustee in bankruptcy is threatening to "assume" the RSS license . . . and assign it to Big Brother. What are RSS's options?

The idea behind bankruptcy is damage control. A business has gone bust, it cannot pay its debts, and now the pieces must be picked up. The insolvent company gets relief from debts it cannot pay, and creditors get paid from whatever assets remain. In the United States, those assets can be sold off in a general liquidation pursuant to Chapter 7 of the Bankruptcy Code, or the business can "reorganize" under Chapter 11 and remain a going concern. Creditors split up the proceeds of liquidation or oversee continued operation of the reorganized business and hope to profit from its future success. Either way, the job of the trustee in bankruptcy is to maximize the value of the debtor's assets for the benefit of its creditors.

One way he can do that is to assign valuable contracts, such as licenses, to third parties willing to pay for them. Many licenses contain provisions intended to thwart this possibility, allowing the licensor to terminate should the licensee declare bankruptcy. Outside the IP licensing context, such clauses do not work. The trustee in bankruptcy has broad authority to assume or assign almost any contract involving ongoing obligations. But the operative word here is "almost." In the United States, the trustee's authority is limited if countervailing law permits the licensor to refuse assignments it does not want. That law, of course, is the principle holding IP licenses to be personal and non-assignable without permission. Bankruptcy courts in effect ignore whatever non-assignment provisions may appear in the license and instead apply this background law. In most cases, a licensor can prevent assignment to someone new. RSS, in other words, will not be forced to accept Big Brother as a substitute licensee.

What about assumption? That is, suppose GeneMachine has given up trying to assign the RSS license, but has filed under Chapter 11 and, as a going concern, wants to keep the license. RSS may object just as strongly to remaining the prisoner of a failed company as it would to getting stuck with Big Brother, preferring instead to choose a new licensee itself. The bankruptcy court will probably respect that decision. But RSS should think twice before jettisoning to stay with GeneMachine, which may have valuable customer relationships that will survive reorganization and production capacity that would take another licensee substantial time to duplicate. Besides, other licensees may be in short supply; whatever the merits of RSS's technology, prospects may figure, they were not enough to save GeneMachine.

What if the situation is reversed, and it is RSS rather than its licensee that has declared bankruptcy? Can the trustee strip GeneMachine of the license and assign it to someone it thinks would generate more income for RSS's creditors? For trademark licenses, the answer is generally yes, but, for other IP licenses, GeneMachine can require the licensor to keep the license in force. Mostly, anyway. The rights granted to GeneMachine will still be enforceable, but not obligations requiring affirmative effort on the part of the bankrupt licensor – infringement protection, training and development, or maintenance obligations, for example. A licensee may keep its basic rights but find itself paying a high royalty for not much in return.

Licensing and standards

As explained in chapter 4, most standards-setting organizations will permit contributors to license on a for-profit basis so long they are reasonable and non-discriminatory about it. But what do these requirements mean? To be "non-discriminatory," a license must be available to all prospective takers – including direct competitors. What terms qualify as "reasonable" is a different matter. The royalty-seeking contributor is likely to receive little guidance from the standards-setting body itself. Such organizations live in fear of an unwelcome tap on the shoulder from antitrust-enforcement authorities, since the likely presence of key industry leaders can conjure up images of a cartel, and therefore usually rely on contributors and the marketplace to sort license terms out. While the occasional willingness to consider proposed license terms can confer valuable legal cover on the technology owner, contributors must ordinarily become familiar with prevailing industry licensing practices and adhere to them to the extent possible. Courts generally extend latitude to a patent owner so long as it has exercised some diligence in discerning and following such terms.

Unfortunately, prevailing practice is an imperfect guidepost. Reliance on market forces to shape permissible license terms has engendered something of a Wild West atmosphere, where demands vary enormously. Let us say that big industry players jointly establish a new packet-handling protocol for wireless data, and contribute it to a standards body. Mindful of the consortium's rules requiring disclosure of relevant IP, one of the contributors confesses to owning an essential patent. It promises the organization that it will license the patent on "non-discriminatory" terms and at a royalty rate of 2.5%. While that seems unambiguous enough, it may turn out that not everyone pays the same license fee. Non-discriminatory does not necessarily mean uniform.

How could that be when the royalty rate is fixed? Well, the initial fee may be 2.5% of the price, say, of the first 10,000 units. Above that, the rate may ratchet down – giving an advantage to the larger firm, but from the patent owner's perspective, encouraging wider use of the standard. Volume discounts of this kind are typically viewed as legitimate if offered to everyone. Licensors may also charge an initial, one-time issue fee for the license – arguably to cover administrative handling costs, but often amounting to many times such costs. Once again, the size of the up-front fee may depend on the identity of the prospective adopter of the standard. And the licensor ordinarily is not obliged to divulge the criteria it applies.

Even a seemingly firm figure like 2.5% can prove slippery. Do companies pay 2.5% of the price of the plug-in interface card that in fact handles the data packets? Or is the relevant baseline the price of the big, expensive box the card is plugged into? Patent owners often take the latter view, leading to huge discrepancies in what companies pay to implement the standard – discrepancies that may be more an accident of product offerings than value actually added to a product. Standards adopters pass along to customers some portion of the royalties they pay, and, if different industry players face different licensing costs, some will obtain a marketplace advantage.

Liberties may also be taken with grantbacks. Those covering innovations tied to the licensed technology – for example, extensions or improvements – are common and usually considered reasonable if they are non-exclusive. But some standards contributors seek more, even insisting on royalty-free access to the adopter's entire IP portfolio. That can be risky, since grantbacks are subject to antitrust scrutiny if they have an anticompetitive effect. In a "network" context where a single licensor accumulates grantbacks from numerous licensees, standards contributors that demand far more than they bestow may be asking for trouble.

An economist might have little sympathy for companies convinced they are paying too much (or giving too much away) in order to implement an industry standard. After all, if enough of the industry feels that way, the standard will simply wither and die. But this view assumes perfect information, which is often in short supply. The standards-setting process is transparent when it comes to technical criteria and, by and large, to the existence of relevant proprietary rights. But the aversion of standards organizations to licensing questions and the veil of secrecy that surrounds most individual licensing agreements impede rational decisionmaking.

Bowing to entreaties from customers, and not knowing if its competitors have adopted the standard or on what terms, a company may feel compelled to go along on whatever license terms it is given. The more industry players that follow suit, the greater will be the network pressure on others to tag along. Perhaps one day standards organizations will begin to require as much transparency in formulating license terms as they do in fashioning the standards themselves – not risking antitrust exposure by agreeing or collaborating on such terms, but at least requiring their disclosure. Then the industry can judge the standard by financial as well as technical criteria, before the snowball effect of growing adoption curtails industry leverage and forces acceptance at virtually any cost. That day, unfortunately, has yet to arrive.

Licenses with universities and research institutions

Since federal policy in the United States began encouraging commercialization of government-funded technology, academic institutions have emerged as critical players in the market for technology. No longer cloistered in their thinking, institutions that once disdained the taint of commercial involvement now relish it – enjoying not only the financial rewards of sponsored research and technology spin-outs, but attracting top-flight faculty who might otherwise pursue careers in the private sector. Academic institutions worldwide have become sophisticated in their understanding of IP and licensing, and in dealing with industry sponsors, technology companies, and investors.

Still, negotiations between a university and outsiders can become mired in provisions that, to the perplexed industry manager, appear legalistic or gratuitous. Royalty terms seldom provoke controversy; academic institutions usually request only a modest percentage of sales. Instead, negotiations may founder on requirements that the university sees as crucial to its academic mission, but which diverge from ordinary commercial concerns.

To deal effectively with universities, therefore, business managers and investors must familiarize themselves with the academy. They must appreciate the goals of academic licensing programs and recognize the players involved. And they must make a determined effort early on in the relationship to develop a shared set of expectations.

Finding a research partner. Businesses usually become acquainted with a university's research capabilities through faculty contacts, e.g., the ties of recently hired graduates to their former advisers or the reputations of eminent professors. Such relationships can lead anywhere from requests for licenses under existing university patents all the way to elaborate, company-funded research programs that culminate in a license only when the project is in large part near completion. University spin-outs, on the other hand, tend to emerge directly from faculty research efforts, often with the encouragement of professional investors eager for an early entrance.

But researchers do not sign deals. Most universities have central offices of technology licensing, staffed by business-oriented professionals familiar with rights transactions. The licensing office, in turn, answers to the university's central administration, which hires its personnel, fixes its budget, and sets general licensing policies. Such policies might include, for example, royalty rates, the willingness to grant exclusive rights to a single company, and how faculty and students share the fruits of commercial success.

But the overall relationship necessarily extends beyond the licensing office, and different university constituencies often have inconsistent priorities. Faculty may seek funding and equipment donations from large corporate sponsors in order to improve their laboratories, whereas the central administration may be more interested in boosting the local economy (and the university's bottom line) by licensing start-ups in exchange for royalties on future products. When a roadblock appears, it is important to determine which constituency has erected it, and who has the power to tear it down.

What the university wants. One shared goal among all research institutions is extending knowledge and enhancing the prestige of academic programs. In general, that means basic research – fresh ways of approaching problems that turn professors into luminaries and even Nobelists. Faculty members almost always receive a share of the university income their work generates, but rarely is profit the overriding motive. To a corporate sponsor, on the other hand, profit is the point. Basic research is fascinating, but corporate investments must, at some point, earn a return. In a spin-out situation, the company and its investors need license terms they can live with – terms that give the start-up a long enough runway of exclusivity to achieve commercial liftoff, and payments tied to revenue-producing events rather than a fixed schedule: milestones, in other words, that are not millstones around the company's corporate neck.

The most important component of university licensing, then, is clear expectations on both sides. That comes easily when the technology has already been developed and patented, in which case a simple license will do. But for deals that include research sponsorship:

- Plan on spending quality time with the principal investigator.
- Specify goals and timing in detail.
- Assess faculty appetites for applied research geared toward the market (as opposed to basic research).
- Ensure that priorities are shared.
- Above all, stage the work so that interim progress can be checked before further funding is committed. This will permit re-evaluation of commercialization potential as the research progresses, and may also give the company some leverage in setting near-term research objectives.

The university's educational mission always figures prominently in the licensing terms it can offer. Corporate research is usually a secretive affair, with results published only to secure proprietary rights (like patents) or to convince customers they should buy. Universities and their professors and students, in contrast, must publish – they cannot put the interests of corporate sponsors above academic obligations. They will, however, usually agree to short-term embargoes on publication, during which company managers can evaluate whether to patent research results. But forget about trade secrets; there are none in academia.

Universities are also sensitive to their public image. Reputations can be tarnished by arrangements that appear to tie up rights for an inordinately long period, denying the public access to fundamental advances. So they tend to resist granting rights to later-developed improvements not directly funded by the licensee, and may insist on limiting license rights to a specific field of use.

Finally, universities stay well out of the line of legal fire, insisting on stringent provisions (including indemnification) limiting institutional liability. They almost always disclaim all warranties, especially of non-infringement. Their licensing offices' discretion to negotiate these clauses is limited – indeed, their terms may be prescribed by law. But, typically, universities will cede control of infringement suits to the licensee.

The supposedly recent trend toward "outsourcing" technology development really began decades ago, with corporate sponsorship of institutional research. Despite the long history, wide cultural gaps and divergent goals still exist – and that is a healthy thing. Independence from marketplace pressures and pursuit of excellence are ultimately what draw industry support. Successful partnerships respect differing priorities without sacrificing shared values.

The reach-through problem

Traditional pain relievers such as aspirin and ibuprofen work by inhibiting two enzymes involved in inflammation: COX-1 and COX-2. It turns out that pain relief requires inhibition only of the COX-2 enzyme. Blocking COX-1 is not only unnecessary but harmful, causing abdominal pain and eventually ulcers. In 1992, scientists at the University of Rochester developed a screening mechanism to determine whether a candidate drug could inhibit the culprit COX-2 inflammatory enzymes without affecting the beneficial COX-1 enzymes. They did not invent any drugs themselves – just a research tool that pharmaceutical companies could use to do that job. We have seen how universities tend to approach licensing, but, in this case, how should they set the price?

From the university's perspective, the number of potential customers is small but the investment that went into development of the tool was substantial; somehow that investment must be recouped through revenue from a limited number of sales. To a pharmaceutical company, however, the tool has value only insofar as it enables successful development of a marketable pain reliever. Because there can be no guarantee such a product will emerge, the tool is, in essence, a lottery ticket.

To tie the price of a research tool more closely to the value it provides, tool developers often resort to so-called "reach-through" royalty arrangements, whereby the price paid by the buyer depends on the revenue, if any, resulting from commercialization of products identified through use of the tool. This allows the developer to charge a relatively small "lottery ticket" price up front, but participate in the customer's upside should the ticket prove a winner. It also accommodates situations in which a prospective customer's hesitation stems less from doubt over the tool's ability to deliver winners than concern over the ultimate cost of bringing products to market. By tying royalties to products actually sold, the research-tool user not only postpones these payments until it earns profits, but may also negotiate a royalty rate that reflects the up-front investment and development costs likely to be expended beforehand.

Reach-through royalty arrangements have attracted widespread attention, much of it negative, due both to their novelty and the vague sense that, once having parted with a product and received its reward, a seller's interest should end. Research-tool sellers beg to differ, and have attempted to formalize their entitlement to reach-through royalties by way of creative patent prosecution and also through licensing. Unfortunately for such sellers, these strategies have generally received a chilly reception.

The University of Rochester hustled its application through the patent office before the reach-through question garnered visibility, and received protection not only on the screening technique, but on as-yet-unknown products identified through its use. Had the patent office given the question sufficient thought, it would have refused the latter claims on the theory that you cannot claim what you have not taught. It did not, though, so the question eventually reached the courts, which invalidated the reach-through claims. Patent-granting authorities worldwide have concurred, and the future for reach-through patents appears bleak indeed.

Faced with this hostile legal climate, research-tool innovators have predictably shifted their focus to contractual approaches. Many have already taken advantage of the broad latitude that parties to a patent license enjoy to shape fee arrangements, basing royalties on the reach-through concept. The question now is whether courts will permit research-tool developers to gain at the negotiating table what cannot be had through the patent system. As explained above, attempts by patent owners to obtain royalties on subject matter outside the scope of their patents usually fall within the concept of patent misuse. But it is precisely those downstream products that will retrospectively determine the value of the research tool. To quarantine the tool's originator from profits resulting from its use would ignore the procompetitive, technology-disseminating effects likely to be generated through such incentives.

Definitive rulings on reach-through royalties are in unfortunately short supply. The National Institutes of Health in the US has criticized the practice, and requires its grant recipients and contractors to refrain from reach-throughs entirely. While it is possible to imagine arrangements that skirt direct reliance on downstream profits but still recognize when value has been derived – for example, staged payment provisions that defer large installments until milestones, such as completion of clinical trials and market introduction of a product, occur – the ultimate price still remains unconnected to the true economic value added.[10] The tool developer will inevitably wind up with too much or, more likely, too little. Hopefully this reality will sink in and reach-through royalties will ultimately become an accepted form of risk sharing.

IP valuation

How much is a license worth? How much, for that matter, is a patent or an entire portfolio worth? IP valuation is a common practice: in mergers and acquisitions, in bankruptcy, in IP litigation, and for tax purposes. But that does not mean there is any consensus on how to do it, or reason to think it can even be done with anything approaching precision.

The task is easily enough stated. An IP-owning firm has earnings. To value its IP assets, you isolate the component of the earnings attributable to the IP, and multiply the fraction it represents by the company's market value. In concept, the exercise is similar to setting a royalty rate, which also focuses on the special profit attributable to IP coverage. Unfortunately, that is where the similarity ends. As evidenced merely by the number of accepted (and often contradictory) methodologies for valuing IP assets, the process involves substantial guesswork. The dirty little secret of IP valuation is that it is often possible to achieve almost any desired outcome based on perfectly plausible assumptions.

The reason is that, outside some unusually well-bounded circumstances, the guesswork begins almost immediately. Most valuation techniques center on a future stream of benefits. Identifying those benefits is easy in the case of a license based on specific IP, having a known term, and generating predictable revenue year to year. We can discount the revenue stream to present value based on the cost of capital, perhaps tweaked to account for the possibility of an adverse litigation outcome that destroys or limits the strength of the licensed IP. The problem is that this paradigm describes almost no real-world

[10] The economically minded reader may object that a hardware store charges one price for a hammer regardless of the fortune a purchaser might derive from its use – that is, whether it is used to build a doghouse or Bill Gates's house – so why should the purveyor of research tools profit from the end product? Underpinning that objection, however, is the assumption that it is the user's efforts with the tool, rather than the tool itself, that dictate the value of the end result. This logic frequently does not apply to use of research tools. In most cases, responsibility for success will fairly be viewed as shared between the tool's developers and its users. Indeed, in some circumstances, use of the tool is likely to be straightforward and rote, perhaps not even requiring technical training.

situations. Most patents are not licensed; rather, they are clutched tightly and used to establish market exclusivity. As we saw in chapter 4, while it is possible to assess qualitatively whether an IP portfolio is providing business value, quantifying that value is another matter. A company's position in the market – even an exclusive position – may derive from its IP, from something else (price, customer service, branding, convenience), or from some unknown and unknowable combination. One can no more attribute a specific portion of revenue to IP than identify the particular egg in a forkful of omelette.

The exercise should be far more straightforward for licensed IP, but even here royalties can fluctuate wildly depending on marketplace acceptance of covered products, the aggressiveness and business skills of the licensee, the evolving strength of the portfolio, the ability of competitors to design around patents, and their commercial incentive to do so. Contrast valuation of a product company versus a licensing company. It may be reasonable to value a product company based on current and recent revenues, on the theory that, as some markets mature, the company's innovation potential will be deployed to open others. A product company's trajectory will not likely differ dramatically from those of other industry players, subject as they all are to the same market forces, so historical growth patterns within the industry segment may reasonably inform the analysis.

Not so in the world of IP. A patent typically covers a specific type of product or, more generously, a particular approach to solving a problem. If the market rejects the patented approach, the patent is worthless; no more innovation remains to be extracted from it. The licensing company may develop other IP, of course, but, to be valuable, any particular patent must clear many hurdles outside the company's direct control – review by the patent office, the production and commercialization capabilities of a licensee, challenge by outsiders. A licensing company's relationship to the marketplace, in other words, is far more attenuated than the product company's, rendering valuations of its IP-grounded asset base that much more speculative.

Let us consider some of the most popular IP valuation techniques and their limitations in light of the above concerns.

Income approaches estimate the value of IP assets based on their effect on enterprise cash flow – that is, by quantifying and discounting to present value the future benefits attributable to the IP (as distinct from other assets). Again, this is akin to unscrambling the proverbial omelette. But, rather than attempting to isolate the IP-derived revenues, appraisers may instead estimate the revenues generated by *everything else* and then subtract this from overall business earnings. This "residual income" approach divides a business into working capital, tangible assets, and intangible assets. Net returns on capital, tangible assets, and non-IP intangible assets are computed based on accepted measures of the liquidity and risk of each asset class. That's *everything else*; the remainder of the company's income is presumed to arise from the IP.

Well, the residual income surely originates somewhere (assuming the figures corresponding to *everything else* are reasonably accurate). But not necessarily from the IP. The residual-income approach merely states a figure – not a mechanism explaining how IP produces that figure. Liquidity and risk measures represent broad averages rather than universal constants accurate for every situation. For example, underestimating the income contribution of a company's tangible assets (due, say, to productivity increases not reflected in the standard measures) overvalues the contribution of IP relative to everything else. Needless to say, residual approaches cannot isolate the values of particular items of IP.

Other income-based approaches attempt to quantify the special profit associated with IP, and suffer from the shortcomings detailed above. The "relief from royalty" technique, for example, values IP assets based on the royalties they generate. But, unless a patent is licensed, how can one identify a revenue stream attributable to it? And, even if it is licensed, who can predict how royalties will vary from year to year, or whether the technology will be eclipsed by new entrants? Sony had high hopes for its Betamax video-recording standard, but, despite its technical superiority and Sony's determined marketing efforts, the standard lost out to VHS. Indeed, one virtue of the residual-income approach is precisely its gross valuation of IP; by not even attempting to discriminate between the exclusionary effects of a company's patents and the branding effects of its trademarks, this "top–down" approach at least avoids combining errors from misguided individual valuations. The "bottom–up" process of building an overall IP value from individual royalty estimates suffers from just this problem.

The difficulty may be more manageable (sometimes, anyway) in the case of trademarks. The "premium profit" approach considers how much additional profit a trademark owner earns from the branded product relative to a generic version. Certainly for consumer products and over-the-counter pharmaceuticals it is straightforward to identify generic equivalents, and, assuming similar cost structures, estimate the excess profit. Even here, though, uncertainties enter the analysis: pricing can reflect market share and distribution considerations in addition to branding effects.

The market approach compares IP assets with similar assets that have been recently licensed or sold by others. One major impediment to successful use of this technique is the lack of public information on such transactions, particularly their key provisions – the price, terms, conditions, and restrictions. Even if enough information can be obtained, however, the very notion of "similar" IP assets remains highly problematic. Patents can be similar in many irrelevant ways, or ways that are relevant in some circumstances but not others. Should patents covering the Betamax standard be considered similar to or comparable with those covering VHS? They certainly involve the same technical field, and maybe they even have comparable scope, but no one would argue that their values even remotely compare.

The cost approach utilizes the historical cost of creating the IP – that is, of obtaining the patent and, in some variations, also including the cost of invention – as a surrogate for IP market value. It is difficult to imagine a more pointless or misleading form of appraisal. Patent costs fall within a relatively narrow band without regard to their ultimate worth, and no one, not even the most insightful business executive, knows in advance what that worth will be. But she can be confident it will have nothing to do with procurement costs, since these stand entirely removed from the market covered by the patent. The price of a patent, in other words, exerts no influence on its value.[11] Considering the costs attributable to the invention likewise sheds little additional light on IP value. While it might seem appealing to assume that rational business people expend research money in lockstep with eventual market value, history proves otherwise.

There are other approaches, some fairly complex,[12] to appraising IP, but fancier footwork does not mitigate the irreducible uncertainties. Valuation is not sophistical or fraudulent unless these uncertainties are brushed aside by an appraisal technique's apparent numeric precision. Using such techniques to derive a defensible fair market value of IP for transfer-tax purposes is one thing; making business decisions based on numbers that represent no more than a guess is something else entirely.

[11] One might argue that low-value patents are not usually filed abroad, and since foreign filing adds significantly to procurement expense, *very* costly (that is, broadly filed) applications tend to reflect the owner's perception of high value. But even here the relationship is speculative, since the decision to file in foreign countries is usually made long before the ultimate value of a patent can be estimated realistically – that is, foreign filing is more a matter of hope than experience.

[12] Option pricing using semi-arcane math seems to be the latest trend.

Index